Yinping Yang

Numerical methods for the inverse dynamics simulation of underactuated mechanical systems

Schriftenreihe des Instituts für Mechanik
Karlsruher Institut für Technologie (KIT)

Band 5

Herausgeber:
Prof. Dr.-Ing. habil. Peter Betsch
Prof. Dr.-Ing. habil. Thomas Seelig

Eine Übersicht aller bisher in dieser Schriftenreihe erschienenen Bände finden Sie am Ende des Buches.

Numerical methods for the inverse dynamics simulation of underactuated mechanical systems

by
Yinping Yang

KIT Scientific Publishing

Dissertation, Karlsruher Institut für Technologie (KIT)
Fakultät für Bauingenieur-, Geo- und Umweltwissenschaften
Tag der mündlichen Prüfung: 06. Oktober 2016
Referenten: Prof. Dr.-Ing. habil. Peter Betsch, Prof. Dr.-Ing. habil. Robert Seifried

Impressum

KIT Scientific Publishing

Karlsruher Institut für Technologie (KIT)
KIT Scientific Publishing
Straße am Forum 2
D-76131 Karlsruhe

KIT Scientific Publishing is a registered trademark of Karlsruhe Institute of Technology. Reprint using the book cover is not allowed.

www.ksp.kit.edu

Print on Demand 2017 – Gedruckt auf FSC-zertifiziertem Papier

ISSN 2363-4936
ISBN 978-3-7315-0626-3
DOI 10.5445/KSP/1000065049

Numerical methods for the inverse dynamics simulation of underactuated mechanical systems

Zur Erlangung des akademischen Grades eines

DOKTOR-INGENIEURS

von der Fakultät für

Bauingenieur-, Geo- und Umweltwissenschaften

des Karlsruher Instituts für Technologie (KIT)

genehmigte

DISSERTATION

von

M.Sc. Yinping Yang

aus Datong, China

Tag der mündlichen Prüfung: 06.10.2016

Hauptreferent: Prof. Dr.-Ing. habil. Peter Betsch
Korreferent: Prof. Dr.-Ing. habil. Robert Seifried

Karlsruhe 2016

Abstract

The present thesis deals with the inverse dynamics simulation of underactuated multibody systems. In particular, the study focuses on solving trajectory tracking control problems of differentially flat underactuated systems. The use of servo constraints provides an approach to formulate trajectory tracking control problems of underacutated systems, which are also called underactuated servo constraint problems. The formulation of underactuated servo constraint problems makes use of minimal coordinates and dependent (or redundant) coordinates to yield a set of differential-algebraic equations (DAEs) with high index. The transition between the redundant coordinates formulation and the minimal coordinates formulation is achieved by applying the discrete null space method. Since the numerical solution to the DAEs with high index is a challenging task and the flatness-based analytical solution is not feasible for complicated underactuated systems, it is necessary to apply index reduction methods to reduce the index before the direct time discretization is performed. A specific projection method is applied to reduce the index from five to three and it requires the computation of projection matrices, which are constant Boolean-type in the redundant coordinates formulation and are time-dependent in the minimal coordinates formulation. A newly proposed index reduction method called index reduction by minimal extension is developed in this thesis and applied to servo constraint problems of underactuated systems. Representative numerical examples are used to demonstrate the application of both index reduction methods. Special attention is placed on the new index reduction by minimal extension method through several advanced examples, which can not be solved by application of the projection method.

Keywords: Inverse dynamics, differential-algeraic equations, trajectory tracking, servo constraints, differential flatness, index reduction, projection method, minimal extension, multibody dynamics, underactuated mechanical systems, feedforward control, cranes, manipulators

Kurzfassung

In der vorliegenden Dissertation wird die Simulation der inversen Dynamik unteraktuierter Mehrkorpersysteme behandelt. Insbesondere werden Steuerungsprobleme der Bahnverfolgung für differentiell flache unteraktuierte Systeme untersucht. Mit Hilfe von Servobindungen werden die Steuerungsprobleme der Bahnverfolgung für unteraktuierte Systeme formuliert. Die betrachteten Probleme werden unteraktuierte Servobindungsprobleme genannt. Minimalkoordinaten, abhängige oder redundante Koordinaten werden zur Formulierung unteraktuierter Servobindungsprobleme verwendet. Die Formulierung ergibt differential-algebraische Gleichungen mit hohem Index. Die diskrete Nullraum-Methode ermöglicht den Übergang von redundanten Koordinaten zu Minimalkoordinaten. Da die numerische Lösung der differential-algebraischen Gleichungen mit hohem Index anspruchsvoll ist und die flachheitsbasierte analytische Lösung für komplizierte unteraktuierte Systeme nicht praktikabel ist, werden Methoden zur Indexreduktion vor der direkten Zeitdiskretisierung eingesetzt. Eine spezielle Projektionsmethode wird angewendet, um den Index von fünf auf drei zu reduzieren. Die Methode erfordert die Berechnung von Projektionsmatrizen, die in der redundanten Koordinaten Formulierung konstant und in der Minimalkoordinaten Formulierung zeitabhängig sind. Eine neue Methode, Indexreduktion durch minimale Erweiterung genannt, wird in dieser Dissertation entwickelt und für Servobindungsprobleme unteraktuierter Systeme verwendet. Die beiden Methoden werden auf repräsentative numerische Beispiele angewandt. Insbesondere wird schon gezeigt, dass sich die neu entwickelte Indexreduktionsmethode zur Lösung involvierter Probleme eignet, die bislang mit der Projektionsmethode nicht gelöst werden konnten.

Schlüsselwörter: Inverse Dynamik, Indexreduktion, Bahnverfolgung, Servobindungen, Mehrkörperdynamik, unteraktuierte Systeme, differential-algebraische Gleichungen, Differentielle Flachheit, Projektionsmethode, minimale Erweiterung, Vorsteuerung, Krane, Manipulatoren

Acknowledgements

This thesis was carried out during my research work from 2010 to 2013 at the Chair of Computational Mechanics at the University of Siegen, and from 2013 to 2016 at the Institute of Mechanics (IFM) at Karlsruhe Institute of Technology (KIT). I would like to thank those people who gave me support and help during this important period in my life.

First and foremost, I would like to express my sincere appreciation and deepest gratitude to my supervisor, Prof. Dr.-Ing. habil. Peter Betsch, for offering me the opportunity to start my career in the academic field and for bringing me into the interesting field of computational mechanics. I am also grateful for his constant encouragement and support and invaluable guidance and suggestions in completing this thesis. Throughout my research study, he has always been patient and encouraging in guiding me to the right direction and solving various problems. I also appreciate all his contributions of time, ideas, and suggestions to this thesis. His encouragement has been a great source of my confidence and motivation. I also would like to thank Prof. Dr.-Ing. habil. Robert Seifried, for his effort being my second supervisor and for his interest in my thesis. I also would like to thank Prof. Dr.-Ing. habil. Thomas Seelig for being my examiner.

I would like to thank Prof. Dr.-Ing. habil. Christian Hesch for his help in computer-related technical problems. Particularly, I appreciate Dr.-Ing. Marlon Franke for his patient answers of my questions and kindly help in the daily life. Many thanks go to all the colleagues at IFM for providing such a wonderful environment for my research.

Most importantly, I would like to give my special thanks to my wife Lihui for her everlasting love, continuous supporting and encouraging. Especially I am very grateful for her help in finding the spelling and grammatical errors and giving valuable contextual proposals. Her love is the driving force of all my inspiration and energy.

Karlsruhe, December 2016 Yinping Yang

Contents

List of Figures

List of Tables

Glossary of notation

Notation	Description
$\boldsymbol{J}$	inertia tensor
$\in$	for all
$\mathbb{R}$	real numbers
Q	configuration manifold
$T_q Q$	tangent space of Q at $\boldsymbol{q}$
$\subset$	is a subset of
$SO(3)$	special orthogonal group
$\boldsymbol{E}$	Euler tensor
g	standard gravity
$\overline{\nabla} f$	discrete derivative of a function f
$\mathbb{I}_n$	a representative time interval

Abbreviation	Description
DAE	differential-algebraic equation
ODE	ordinary differential equation
BEM	basic energy momentum scheme
EM	energy momentum scheme
VI	variational integrator
Gen-α	generalized-α scheme

1 Introduction

Inverse dynamics problems arise in many applications of feedforward control problems, such as robotic control, aircraft control or crane control. The present thesis deals with inverse dynamics simulation problems and especially focuses on solving trajectory tracking control problems. The goal of trajectory tracking control problems is to determine control inputs that force a mechanical system to complete a prescribed motion. The determination of control inputs is based on the dynamic model of the controlled mechanical system, which can be formulated by employing either generalized coordinates or redundant coordinates.

The formulation of inverse dynamics simulation problems yields differential-algebraic equations (DAEs), because the desired system outputs expressed in terms of state variables lead to servo constraints on the system. If fully actuated systems are considered, the number of control inputs/outputs is equal to the number of degrees of freedom. In fully actuated systems control inputs are easily solved from the dynamic equations by the routine inverse dynamics analysis, since the system motion is fully specified by the task requirements. Besides, the governing equations for servo constraint problems of fully actuated systems are (differentiation) index-3 DAEs. The servo constraint problem of fully actuated systems is well understood in the application of robot control. For example, the inverse dynamics control of such problems can be used to generate manipulator control torques.

In contrast to fully actuated systems, the situation changes considerably for underactuated systems in which the number of control inputs/outputs is lower than the number of degrees of freedom. The system motion is specified by desired system outputs with the same number as control inputs. Due to the property of underactuation, the inverse dynamics simulation of underactuated systems is much more demanding. Control inputs can not be solved from the dynamic equations by model inversion, since the input distribution matrix in the governing equations is not invertible. Therefore, the determination of control inputs that force the underactuated system to complete the partly specified motion is a challenging task. In particular, the governing equations for servo constraint problems of underactuated systems arise as a set of DAEs with high index. Servo constraint problems of underactuated systems in partly specified motion can be viewed from the perspective of constrained motion. From the geometric viewpoint of Blajer [27], similar to geometric constraint forces in constrained multibody systems, control inputs (actuator forces) can be regarded as reaction forces of servo constraints. However, the reactions of servo constraints may have any direction with respect to the manifold of servo constraints, and in the extreme case may be tangent, while the reactions of geometric constraints are orthogonal to the respective constraint manifold. The realization of servo constraints with the use of control forces can range from orthogonal to tangential. In the case of tangential realization, underactuated systems are differentially flat. The solvability of the DAEs with high index for differentially flat underactuated systems (the controllability of differentially flat underactuated systems in partly specified motions) is closely related to the mathematical property known as differential flatness [39], which implies that all system state variables and control inputs can be algebraically expressed in terms of desired outputs and their time derivatives up to a certain order, without integrating any differential equations. However, the flatness-based analytical approach is not feasible for more complicated underactuated systems and the derivations are featured by substantial complexity. The DAE formulation provides a more convenient appraoch to the inverse dynamics analysis of underactuated systems in partly specified motion. Servo constraint problems of differentially flat underactuated systems arise in many applications, such as control of cranes, control of robots with flexible joints or flexible members and control of cable suspension manipulators.

In the formulation of underactuated multibody systems, the choice of coordinates has strong impact on the form of the equations of motion. If generalized (minimal) coordinates are used, the governing equations of motion are in the form of ordinary differential equations (ODEs), which are in general highly nonlinear. The orientation of rigid bodies is described by rotational parameters such as Euler angles, which give rise to the singularity problem. By contrast, if redundant coordinates are applied, the formulation of constrained mechanical systems yields differential-algebraic equations (DAEs), which exhibit a comparatively simple structure. The description of the orientation of rigid bodies relies on direction cosines instead of rotational parameters such as Euler angles, rotation vectors or other 3-parameter description of finite rotations. Thus, the rotationless formulation is featured by a constant mass matrix and can be easily extended to flexible multibody dynamics.

In the present thesis, numerical methods are developed to deal with servo constraint problems of underactuated multibody systems. In particular, the study focuses on differentially flat underactuated mechanical systems, in which the index of the DAEs exeeds three. The high index value causes difficulties in the direct numerical integration of the DAEs. Therefore, to facilitate a stable numerical integration, index reduction methods are preferred to reduce the high index value of the DAEs to three or even lower. A specific projection method proposed by Blajer and Kołodziejczyk [27] yields an index reduction from five to three. The projection method requires the computation of projection matrices, which are time-dependent for the formulation in terms of generalized coordinates and constant Boolean-type for the formulation in terms of redundant coordinates. The purpose of the projection matrices is to split the dynamics of the underactuated system into constrained and unconstrained parts. After the application of the projection method, the high index DAEs are transformed to a more tractable index-3 form, which is amenable to a direct discretization with common numerical integration schemes such as the backward Euler method. An alternative index reduction method is the newly proposed method to reduce the index of the DAEs. The new approach relies on the index reduction by minimal extension originally developed by Kunkel and Mehrmann [62] for more general DAEs. The technique of minimal extension is especially suited for the semi-explicit structure of the DAEs and it is not necessary to compute projection matrices as in the projection method. Therefore, the new method, also called index reduction by minimal extension, can be applied to servo constraint problems of underactuated systems to reduce the index value of the DAEs to three or even to one. Moreover, the reduced index-1 DAEs are purely algebraic and reflect the fact that the underactuated system at hand is differentially flat. At last, the desired control inputs that force the underactuated system to complete the partly specified motion are determined by solving the resulting DAEs and the feedforward control strategy is obtained for the trajectory tracking control of underactuated systems.

1.1 Literature review

A bief literature review on servo constraint problems of underactuated systems is given below.

In servo constraint problems, control outputs (specified in time load coordinates) lead to servo [56] (control [76] or program [25]) constraints. The formulation of underactuated servo constraint problems is accomplished by employing either minimal coordinates [28, 20, 21, 94, 2] or redundant coordinates [20, 21, 30, 31, 94, 2]. The rotationless formulation in terms of redundant coordinates is widely employed in diverse numerical problems, such as the numerical integration for constrained rigid multibody dynamics [17, 9, 13, 19, 87], for flexible multibody dynamics [14, 15, 79], for optimal control problems [23, 84], for contact problems of flexible bodies [12, 51, 41, 40], and for nonlinear thermo-viscoelastodynamics problems [47, 48, 49, 61, 60]. The present rotationless formulation has similarities to the formulation based on natural coordinates advocated by García de Jalón et al. [37, 38]. It can be reduced to the formulation in terms of generalized coordinates by application of the discrete null space method with nodal reparametrization [9, 13, 67, 87]. In addition, a specific coordinate augmentation technique [19, 88, 87, 40] is applied in the rotationless formulation to incorporate rotational variables and associated torques. The formulation of underactuated servo constraint problems yields equations of motion in the form of DAEs with high index, which are difficult to be treated in the numerical integration. Therefore, index reduction methods are applied to transform the high index DAEs to lower index DAEs, which are amenable to a direct discretization. Blajer and Kołodziejczyk [25, 27] have originally proposed a specific projection method to deal with servo constraint problems. In particular, two projected formulations are distinguished in the application of the projection method. These are the projected formulation in terms of minimal coordinates [28, 29, 21, 94] and the projected formulation in terms of redundant coordinates [30, 31, 21, 94]. A new index reduction method is called index reduction by minimal extension, which was originally developed by Kunkel and Mehrmann [62, 63]. The new index reduction method has been applied to the servo constraint problems of underactuated systems in [2, 24], to the inverse dynamics simulation [95, 96] of a class of cranes [65, 59, 57], whose formulation fits into the general framework [58], and to servo constraint problems of kinematically undetermined cable suspension manipulators [53, 70, 52]. Differential flatness [39, 78] is one important mathematical property for differentially flat underactuated systems and yields the flatness-based solution. However, it is not possible to derive the flatness-based analytical solution for complicated servo constraint problems. Numerical methods are therefore much more preferable for solving the problem.

1.2 Outline

This section gives an overview over the thesis and the main contents of each chapter. Chapter 2 provides the basic theoretical concepts for the modeling of rigid multibody systems. It first introduces Hamilton's principle, which is used to derive the Lagrange's equations of the second and first kind. Then the application of Lagrange's equations yields equations of motion for discrete rigid multibody systems under consideration. The governing equations take the form of ordinary differential equations (ODEs) in the generalized coordinates formulation or the form of differential-algeraic equations (DAEs) in the redundant coordinates formulation. In particular, the discrete null space method can be applied to reduce the large number of equations and unknowns present in the rotationless formulation. Therefore, the generalized coordinates formulation can be derived from the redundant coordinates formulation as well. In addition, the descriptions of rigid bodies and two basic kinematic pairs are given in the case of the rotationless formulation. The coordinate augmentation is also shortly introduced. To this end, the numerical example of a three-dimensional rotary crane demonstrates the comparison of two different formulations of rigid multibody systems and their corresponding influences on the numerical integration.

Chapter 3 presents several numerical time-stepping schemes, such as the implicit Euler method, the mid-point-type rule, the energy-momentum scheme and the variational integrator. These methods can be used for the direct time discretization of the DAEs for constrained multibody systems. In this connection, a classical benchmark problem for rigid multibody dynamics is used to demonstrate the application of different numerical integration schemes and the numerical results are discussed and compared with each other.

Chapter 4 describes the inverse dynamics simulation problems of underactuated mechanical systems, which are formulated by using servo constraints. The formulation of servo constraint problems makes use of either generalized coordinates or redundant coordinates. In the case of redundant coordinates, the discrete null space method can be applied to yield the size reduction. The formulation of underactuated servo constraint problems yields differential-algebraic equations (DAEs) with high index. As a specific index reduction method, the projection method is applied to yield the index reduction in order to solve servo constraint problems. In this connection, the projected formulation is used for the formulation in terms of generalized coordinates, redundant coordinates and dependent coordinates, respectively. In addition, the property of differential flatness is introduced and used to derive the analytical reference solution for differentially flat underactuated systems. Two numerical examples are analyzed and demonstrate the comparison between different projected formulations and the flatness-based approach.

Chapter 5 introduces a new index reduction method, which is called index reduction by minimal extension. The method is first described for constrained mechanical systems. Then it is applied to the inverse dynamics simulation of cranes, which are formulated by using dependent coordinates. In this case, index reduction by minimal extension can be applied once more and DAEs of index 1 can be obtained. This proves that the cranes under consideration are differentially flat. Besides, index reduction by minimal extension can also be applied to the formulation of cranes in terms of minimal coordinates. In this case, the procedure is more complicated than in the dependent coordinates formulation. The commutative process between minimal extension and size reduction is proved and shown in a diagram. This chapter introduces also a general formulation in terms of redundant coordinates, in which the number of holonomic constraints is greater than the number of servo constraints. It is proved that index reduction by minimal extension can be applied to this formulation as well. The number of redundant coordinates can be reduced by application of the discrete null space method. For simplicity, the backward Euler scheme is applied for the time discretization of the resulting index-3 DAEs. Two numerical examples show the application of index reduction by minimal extension to the formulation in terms of dependent coordinates and minimal coordinates, respectively. The third example demonstrates the application of index reduction by minimal extension to the formulation in terms of redundant coordinates and the size reduction procedure. The results are presented and compared with the analytical reference solution.

Chapter 6 gives several advanced numerical examples of underactuated mechanical systems and demonstrates the effective application of index reduction by minimal extension to the formulation in terms of redundant coordinates.

Eventually conclusions are drawn and an outlook for future developments is provided in Chapter 7.

2 Modeling of rigid multibody dynamics

Multibody systems are composed of interconnected rigid or flexible bodies that perform translational or rotational motions, and the motion of the bodies is constrained by different types of joints [83]. In computational multibody dynamics, different formulations are used to study the dynamic behaviour of multibody systems. The formulation of multibody systems depends on the choice of coordinates for the description of multibody dynamics and the choice of coordinates also has strong impact on the form of equations of motion.

There will be two alternative formulations introduced in this chapter: the formulation in terms of generalized (or minimal) coordinates along with Euler (or Bryant) angles for the description of the orientation of rigid bodies, and the formulation in terms of redundant coordinates, in which the orientation of rigid bodies is described in terms of direction cosines (see, for example, [9, 13]). The formulation in terms of generalized coordinates yields equations of motion in the form of ordinary differential equations (ODEs). In contrast, the formulation in terms of redundant coordinates yields differential-algebraic equations (DAEs). In addition, a numerical example of a three-dimensional rotary crane will be presented, which demonstrates the application of the two formulations and their influences on the numerical time integration.

2.1 Hamilton's principle

In analytical mechanics one important variational principle is Hamilton's principle [43, 64], from which some fundamental laws of mechanics, like Lagrange's equations and Hamilton's equations, can be derived. Hamilton's principle is an integral principle, which considers the motion of an entire system between two time points t_1 and t_2. It reduces the problem of dynamics to the investigation of a scalar definite integral and has the remarkable advantage of being invariant to the coordinate system used to describe the Lagrangian [74].

The action (or action integral) is defined as

$$S = \int_{t_1}^{t_2} L(\boldsymbol{q}, \dot{\boldsymbol{q}})\,\mathrm{d}t \tag{2.1}$$

where the Lagrangian L is expressed in the form

$$L = T - V \tag{2.2}$$

with the system kinetic energy T and the system potential energy V.

Hamilton's principle can be stated as follows: *The actual path in the configuration space renders the value of the definite integral* $S = \int_{t_1}^{t_2} L(\boldsymbol{q}, \dot{\boldsymbol{q}})\,\mathrm{d}t$ *stationary with respect to all arbitrary variations of the path between two instants* t_1 *and* t_2 *provided that the path variations vanish at these two end points* [74]. Mathematically it implies that the motion of a dynamical system is such that the *variation* of the line integral S for fixed t_1 and t_2 is zero [43]:

$$\delta S = \delta \int_{t_1}^{t_2} L(\boldsymbol{q}, \dot{\boldsymbol{q}})\,\mathrm{d}t = 0 \tag{2.3}$$

2.1.1 Lagrange's equations of the second kind

In rigid multibody systems Lagrange's equations can be derived from Hamilton's principle by using either generalized coordinates or redundant coordinates. Generalized coordinates $\boldsymbol{\mu} \in \mathbb{R}^f$ can be used to describe the configuration of the mechanical system and the dimension f is equal to the degrees of freedom of the mechanical system.

For mechanical systems, if generalized coordinates $\boldsymbol{\mu} \in \mathbb{R}^f$ are used, the variation of the action integral reads

$$\begin{aligned}\delta S &= \delta \int_{t_1}^{t_2} L(\boldsymbol{\mu}, \dot{\boldsymbol{\mu}})\,\mathrm{d}t = \int_{t_1}^{t_2} \delta L(\boldsymbol{\mu}, \dot{\boldsymbol{\mu}})\,\mathrm{d}t \\ &= \int_{t_1}^{t_2} \left[\frac{\partial L}{\partial \boldsymbol{\mu}} \cdot \delta\boldsymbol{\mu} + \frac{\partial L}{\partial \dot{\boldsymbol{\mu}}} \cdot \delta\dot{\boldsymbol{\mu}}\right] \mathrm{d}t \\ &= \int_{t_1}^{t_2} \left[\frac{\partial L}{\partial \boldsymbol{\mu}} - \frac{\mathrm{d}}{\mathrm{d}t}\left(\frac{\partial L}{\partial \dot{\boldsymbol{\mu}}}\right)\right] \cdot \delta\boldsymbol{\mu}\,\mathrm{d}t\end{aligned} \tag{2.4}$$

where the integration by parts is expressed as

$$\int_{t_1}^{t_2} \frac{\partial L}{\partial \dot{\boldsymbol{\mu}}} \cdot \delta\dot{\boldsymbol{\mu}}\,\mathrm{d}t = -\int_{t_1}^{t_2} \frac{\mathrm{d}}{\mathrm{d}t}\left(\frac{\partial L}{\partial \dot{\boldsymbol{\mu}}}\right) \cdot \delta\boldsymbol{\mu}\,\mathrm{d}t + \left[\frac{\partial L}{\partial \dot{\boldsymbol{\mu}}} \cdot \delta\boldsymbol{\mu}\right]_{t_1}^{t_2} \tag{2.5}$$

with the endpoint conditions $\delta\boldsymbol{\mu}(t_1) = \delta\boldsymbol{\mu}(t_2) = 0$. Requiring that the variations of the action integral be zero for all $\delta\boldsymbol{\mu}$ implies that the integrand must be zero for all time t, giving the well-known Euler-Lagrange equations for conservative holonomic systems [71],

$$\frac{\mathrm{d}}{\mathrm{d}t}\left(\frac{\partial L}{\partial \dot{\boldsymbol{\mu}}}\right) - \frac{\partial L}{\partial \boldsymbol{\mu}} = \mathbf{0} \tag{2.6}$$

which are also called Lagrange's equations of the second kind. Based on Lagrange-d'Alembert's principle, for non-conservative holonomic systems there is

$$\delta S = \delta \int_{t_1}^{t_2} L(\boldsymbol{\mu}, \dot{\boldsymbol{\mu}})\,\mathrm{d}t + \int_{t_1}^{t_2} \boldsymbol{Q}^* \cdot \delta\boldsymbol{\mu}\,\mathrm{d}t = 0 \tag{2.7}$$

which leads to Euler-Lagrange equations for non-conservative holonomic systems,

$$\frac{\mathrm{d}}{\mathrm{d}t}\left(\frac{\partial L}{\partial \dot{\boldsymbol{\mu}}}\right) - \frac{\partial L}{\partial \boldsymbol{\mu}} = \boldsymbol{Q}^* \tag{2.8}$$

where $\boldsymbol{Q}^*$ are non-conservative generalized forces, and more specifically, the applied forces that can not be derived from a potential, such as the friction force and the actuation force.

2.1.2 Lagrange's equations of the first kind

If redundant coordinates $\boldsymbol{q} \in \mathbb{R}^n$ are used, the coordinates are not independent, and then constraint equations are required to restrict the motion of the mechanical system and constrain the mechanical system to a lower dimensional manifold.

In the case of holonomic mechanical systems, the constraints can be expressed as constraint functions

$$\boldsymbol{\Phi}(\boldsymbol{q}, t) = \boldsymbol{0} \tag{2.9}$$

where $\boldsymbol{\Phi} \in \mathbb{R}^m$ and the dimension m denotes the number of independent holonomic constraints. If redundant coordinates $\boldsymbol{q} \in \mathbb{R}^n$ have the dimension n, the degrees of freedom of the mechanical system can be calculated by $f = n - m$, which is equal to the number of generalized coordinates $\boldsymbol{\mu} \in \mathbb{R}^f$.

If the constraints can not be expressed in the form of Equation (2.9), the constraints are nonholonomic. If constraint equations are not explicitly dependent on time, the constraints are scleronomic and can be expressed as

$$\boldsymbol{\Phi}(\boldsymbol{q}) = \boldsymbol{0} \tag{2.10}$$

Otherwise the constraints are rheonomic.

For constrained mechanical systems with holonomic constraints, the action integral needs to be modified as

$$\widetilde{S} = S - \int_{t_1}^{t_2} \boldsymbol{\lambda} \cdot \boldsymbol{\Phi}(\boldsymbol{q}) \, \mathrm{d}t = \int_{t_1}^{t_2} \left(L(\boldsymbol{q}, \dot{\boldsymbol{q}}) - \boldsymbol{\lambda} \cdot \boldsymbol{\Phi}(\boldsymbol{q}) \right) \mathrm{d}t \tag{2.11}$$

with Lagrange multipliers $\boldsymbol{\lambda} \in \mathbb{R}^m$. Then the variation of the modified action reads

$$\begin{aligned} \delta \widetilde{S} &= \int_{t_1}^{t_2} \left[\frac{\partial L}{\partial \boldsymbol{q}} \cdot \delta \boldsymbol{q} + \frac{\partial L}{\partial \dot{\boldsymbol{q}}} \cdot \delta \dot{\boldsymbol{q}} - \boldsymbol{\lambda} \cdot \frac{\partial \boldsymbol{\Phi}}{\partial \boldsymbol{q}} \delta \boldsymbol{q} - \delta \boldsymbol{\lambda} \cdot \boldsymbol{\Phi} \right] \mathrm{d}t \\ &= \int_{t_1}^{t_2} \left[\frac{\partial L}{\partial \boldsymbol{q}} - \frac{\mathrm{d}}{\mathrm{d}t} \left(\frac{\partial L}{\partial \dot{\boldsymbol{q}}} \right) - \left(\frac{\partial \boldsymbol{\Phi}}{\partial \boldsymbol{q}} \right)^T \boldsymbol{\lambda} \right] \cdot \delta \boldsymbol{q} \, \mathrm{d}t - \int_{t_1}^{t_2} \boldsymbol{\Phi} \cdot \delta \boldsymbol{\lambda} \, \mathrm{d}t \end{aligned} \tag{2.12}$$

where the integration by parts

$$\int_{t_1}^{t_2} \frac{\partial L}{\partial \dot{q}} \cdot \delta \dot{q} \, \mathrm{d}t = - \int_{t_1}^{t_2} \frac{\mathrm{d}}{\mathrm{d}t} \left(\frac{\partial L}{\partial \dot{q}} \right) \cdot \delta q \, \mathrm{d}t + \left[\frac{\partial L}{\partial \dot{q}} \cdot \delta q \right]_{t_1}^{t_2} \tag{2.13}$$

is used with the endpoint conditions $\delta q(t_1) = \delta q(t_2) = 0$. Hamilton's principle implies that the variation of the modified action integral in the actual path followed by the system must be equal to zero. Then according to the fundamental lemma of the calculus of variations, the coefficients of δq and $\delta \lambda$ in Equation (2.12) are both equal to zero. For conservative holonomic systems this leads to the Euler-Lagrange equations

$$\frac{\mathrm{d}}{\mathrm{d}t} \left(\frac{\partial L}{\partial \dot{q}} \right) - \frac{\partial L}{\partial q} + \left(\frac{\partial \boldsymbol{\Phi}}{\partial q} \right)^T \lambda = \mathbf{0} \tag{2.14a}$$

$$\boldsymbol{\Phi}(q) = \mathbf{0} \tag{2.14b}$$

which are also called Lagrange's equations of the first kind. Analogously, for non-conservative holonomic mechanical systems, the Euler-Lagrange equations read

$$\frac{\mathrm{d}}{\mathrm{d}t} \left(\frac{\partial L}{\partial \dot{q}} \right) - \frac{\partial L}{\partial q} + \left(\frac{\partial \boldsymbol{\Phi}}{\partial q} \right)^T \lambda = Q^* \tag{2.15a}$$

$$\boldsymbol{\Phi}(q) = \mathbf{0} \tag{2.15b}$$

2.2 Generalized coordinates formulation

The generalized coordinates formulation can be employed to model the dynamics of mechanical systems. In this formulation the Lagrange's equations of the second kind in Equation (2.6) (see also [46]) is applied to derive the equations of motion for conservative constrained mechanical systems. In Equation (2.2) the kinetic energy T and potential energy V of a mechanical system need to be determined. The total kinetic energy of the system is the sum of all bodies' translational and rotational energy parts:

$$T = \sum_{I=1}^{N} \left(\frac{1}{2} m_I v_I \cdot v_I + \frac{1}{2} \omega_I \cdot J_I \omega_I \right) \tag{2.16}$$

Here, the mass of the body is denoted by m_I and the velocity of the center of mass of the body is specified by v_I. The angular velocity of the body about an axis is given by

ω_I and the inertia tensor of the body is represented by $\boldsymbol{J}_I$. For the orthogonal principal axis, the products of inertia are equal to zero, and then the inertia tensor is reduced to a diagonal matrix. Besides, the total number of bodies is N. The calculation of the potential energy V is simple and will not be discussed here.

Then, inserting the derived Lagrangian into Equation (2.6) leads to the equations of motion of mechanical systems in the form of ODEs:

$$\dot{\boldsymbol{\mu}} = \boldsymbol{v} \tag{2.17a}$$

$$\boldsymbol{\mathcal{M}}(\boldsymbol{\mu})\dot{\boldsymbol{v}} = \boldsymbol{f}(\boldsymbol{\mu}, \dot{\boldsymbol{\mu}}) \tag{2.17b}$$

with generalized coordinates $\boldsymbol{\mu} \in \mathbb{R}^f$, generalized velocities $\boldsymbol{v} \in \mathbb{R}^f$, the positive-definite mass matrix $\boldsymbol{\mathcal{M}}(\boldsymbol{\mu}) \in \mathbb{R}^{f,f}$, generalized dynamic (centrifugal, Coriolis and gyroscopic) and applied forces $\boldsymbol{f} \in \mathbb{R}^f$. However, the equations of motion in terms of rotational degrees of freedom (for example, Euler angles) are highly nonlinear and their derivation is quite cumbersome. In particular, the mass matrix $\boldsymbol{\mathcal{M}}(\boldsymbol{\mu})$ is configuration-dependent and very complicated.

2.3 Rotationless formulation

In contrast to the generalized coordinates formulation, the rotationless formulation relies on direction cosines and makes use of redundant coordinates to circumvent the use of rotational parameters such as Euler angles, joint angles, rotation vectors or quaternions [16] for the description of the orientation of rigid bodies. In this section the Lagrange's equations of the first kind will be applied to derive the equations of motion in terms of redundant coordinates for conservative constrained mechanical systems. Then the reduced formulation of the DAEs will be introduced and the rotationless formulation will be described in detail.

2.3.1 Equations of motion

In the application of the Lagrange's equations of the first kind, the kinetic energy of the finite-dimensional mechanical system can be given by

$$T(\dot{\boldsymbol{q}}) = \frac{1}{2}\dot{\boldsymbol{q}} \cdot \boldsymbol{M}\dot{\boldsymbol{q}} \tag{2.18}$$

Here, $\boldsymbol{M} \in \mathbb{R}^{n \times n}$ is a constant and symmetric positive-definite mass matrix and the constant mass matrix is an essential property of the rotationless formulation. As before, the superposed dot denotes the time differentiation. The potential energy is given by a function $V(\boldsymbol{q}) \in \mathbb{R}$. Then the equations (2.14a)–(2.14b) are used to derive the dynamic equations for mechanical systems subject to holonomic constraints. The terms in the equations (2.14a)–(2.14b) are determined by

$$\frac{\mathrm{d}}{\mathrm{d}t}\left(\frac{\partial L}{\partial \dot{\boldsymbol{q}}}\right) = \boldsymbol{M}\ddot{\boldsymbol{q}} \tag{2.19a}$$

$$\frac{\partial L}{\partial \boldsymbol{q}} = -\nabla V(\boldsymbol{q}) \tag{2.19b}$$

Then the equations of motion for conservative constrained mechanical systems can be written in the form of differential-algebraic equations (DAEs):

$$\dot{\boldsymbol{q}} - \boldsymbol{v} = \boldsymbol{0} \tag{2.20a}$$

$$\boldsymbol{M}\dot{\boldsymbol{v}} + \nabla V(\boldsymbol{q}) + \boldsymbol{G}^T(\boldsymbol{q})\boldsymbol{\lambda} = \boldsymbol{0} \tag{2.20b}$$

$$\boldsymbol{\Phi}(\boldsymbol{q}) = \boldsymbol{0} \tag{2.20c}$$

Here, a vector of redundant coordinates is given by $\boldsymbol{q} \in \mathbb{R}^n$ which specifies the configuration of the mechanical system. A vector of redundant velocities is denoted by $\boldsymbol{v} \in \mathbb{R}^n$. Moreover, a vector of holonomic constraint functions is expressed by $\boldsymbol{\Phi}(\boldsymbol{q}) \in \mathbb{R}^m$ with the corresponding constraint Jacobian matrix $\boldsymbol{G}(\boldsymbol{q}) = D\boldsymbol{\Phi}(\boldsymbol{q}) \in \mathbb{R}^{m \times n}$ and a vector of Lagrange multipliers is given by $\boldsymbol{\lambda} \in \mathbb{R}^m$ which specifies the relative magnitude of constraint forces. In addition, the m constraints are assumed to be independent. Consequently, the constraint Jacobian matrix $\boldsymbol{G}$ has full row rank.

Due to the holonomic constraints in Equation (2.20c), the $n - m$ dimensional configuration manifold of the mechanical system can be expressed by

$$Q = \{\boldsymbol{q}(t) \in \mathbb{R}^n \mid \boldsymbol{\Phi}(\boldsymbol{q}) = \boldsymbol{0}\} \tag{2.21}$$

Accordingly, the degrees of freedom of the discrete mechanical system is $f = n - m$. In summary, the equations (2.20a)–(2.20c) are index-3 differential-algebraic equations (see [35, 63]), which consist of $2n + m$ equations for the unknown variable vectors $(\boldsymbol{q}, \boldsymbol{v}, \boldsymbol{\lambda}) \in \mathbb{R}^n \times \mathbb{R}^n \times \mathbb{R}^m$. Obviously the DAEs (2.20a)–(2.20c) exhibit a comparatively simple structure, which makes possible the design of energy-momentum conserving schemes (see Section 3.3). Some time-stepping schemes will be explained in the following chapter.

2.3.2 Reduced formulation of the DAEs

The DAEs (2.20a)–(2.20c) can be reformulated by premultiplying an appropriate matrix (i.e. the null space matrix) and reparameterizing the configuration manifold Q. This size-reduction procedure (i.e. discrete null space method) has been dealt with in several works (see [9, 13, 19, 84, 87]). Due to the holonomic constraints (2.20c), redundant velocities are restricted to the tangent space $T_qQ \subset \mathbb{R}^n$. The geometric constraint equations (2.20c) can be differentiated once with respect to time to obtain the constraint conditions at the velocity level. Accordingly, the kinematic constraints assume the form

$$\boldsymbol{G}(\boldsymbol{q})\boldsymbol{v} = \boldsymbol{0} \tag{2.22}$$

and the tangent space at $\boldsymbol{q} \in Q$ can be identified with the null space of the constraint Jacobian, that means

$$T_qQ = \text{null}(\boldsymbol{G}(\boldsymbol{q})) \tag{2.23}$$

Suppose that there exists a matrix $\boldsymbol{P}(\boldsymbol{q}) \in \mathbb{R}^{n\times f}$, whose column vectors span the null space of $\boldsymbol{G}(\boldsymbol{q}) \in \mathbb{R}^{m\times n}$. The matrix $\boldsymbol{P}(\boldsymbol{q})$ is the null space matrix, which satisfies the mathmatical condition

$$\text{range}(\boldsymbol{P}(\boldsymbol{q})) = \text{null}(\boldsymbol{G}(\boldsymbol{q})) \tag{2.24}$$

or in the alternative form

$$\boldsymbol{G}\boldsymbol{P} = \boldsymbol{0} \tag{2.25}$$

Then admissible velocities $\boldsymbol{v} \in T_qQ$ can be written as

$$\boldsymbol{v} = \boldsymbol{P}\boldsymbol{\nu} \tag{2.26}$$

with independent generalized velocities $\boldsymbol{\nu} \in \mathbb{R}^f$. These velocities $\boldsymbol{\nu}$ may be classified as quasi-velocities because their time integrals do not result in generalized coordinates [13]. Equation (2.26) shows that the null space matrix $\boldsymbol{P}(\boldsymbol{q})$ maps $\mathbb{R}^f$ into T_qQ. After inserting Equation (2.26) into (2.18), the reduced form of the kinetic energy $\widetilde{T}$ is expressed as

$$\widetilde{T}(\boldsymbol{q},\boldsymbol{\nu}) = \frac{1}{2}\boldsymbol{\nu}\cdot\widetilde{\boldsymbol{M}}\boldsymbol{\nu} \tag{2.27}$$

where the reduced mass matrix $\widetilde{\boldsymbol{M}}$ is given by

$$\widetilde{\boldsymbol{M}} = \boldsymbol{P}^T\boldsymbol{M}\boldsymbol{P} \tag{2.28}$$

and coincides with the configuration-dependent mass matrix $\boldsymbol{\mathcal{M}}$ in Equation (2.17b).

In order to eliminate the constraint forces, Equation (2.20b) is premultiplied by $\boldsymbol{P}^T$, which means the equations in (2.20b) are projected onto the tangent space T_qQ. Then using the equations (2.25) and (2.26) leads to the reduced formulation of the DAEs:

$$\dot{\boldsymbol{q}} - \boldsymbol{P}\boldsymbol{v} = \boldsymbol{0} \tag{2.29a}$$

$$\widetilde{\boldsymbol{M}}\dot{\boldsymbol{v}} + \boldsymbol{P}^T\boldsymbol{M}\dot{\boldsymbol{P}}\boldsymbol{v} + \boldsymbol{P}^T\nabla V(\boldsymbol{q}) = \boldsymbol{0} \tag{2.29b}$$

$$\boldsymbol{\Phi}(\boldsymbol{q}) = \boldsymbol{0} \tag{2.29c}$$

which govern the motion of the constrained mechanical system. By introducing appropriate generalized coordinates or local coordinates $\boldsymbol{\mu} \in U \subset \mathbb{R}^f$, a second size-reduction can be performed for the parameterization of the configuration manifold Q. Accordingly, the holonomic constraints (2.29c) can be eliminated if a mapping $\boldsymbol{F} : \mathbb{R}^f \mapsto \mathbb{R}^n$ can be found such that

$$\boldsymbol{q} = \boldsymbol{F}(\boldsymbol{\mu}) \tag{2.30}$$

Then the constraints

$$\boldsymbol{\Phi}(\boldsymbol{F}(\boldsymbol{\mu})) = \boldsymbol{0} \tag{2.31}$$

vanish from Equation (2.29c). It should be noted that the differentiation of Equation (2.30) with respect to time gives rise to the consistency condition (2.26). The null space matrix $\boldsymbol{P}$ is then calculated by

$$\boldsymbol{P} = D\boldsymbol{F}(\boldsymbol{\mu}) \tag{2.32}$$

Generally the size-reduction procedure leads to equations of motion (2.17a)–(2.17b), which have already been derived by applying the Lagrange's equations of the second kind in the generalized coordinates formulation.

2.3.3 Spatial rigid body

The specific rotationless formulation [17] of rigid bodies fits into the framework for constrained mechanical systems, makes use of redundant coordinates and circumvents the use of any type of rotational parameters. Thus, it can be employed to describe the position and orientation of a spatial rigid body.

In the present rotationless formulation, the orientation (rotation) of the rigid body is characterized by nine redundant coordinates $\boldsymbol{d}_i \in \mathbb{R}^3$, $(i = 1, 2, 3)$ (see Fig. 2.1), which constitute the rotation matrix (direction cosine matrix). The nine redundant coordinates are subject to six independent holonomic constraints (2.44), sometimes called the internal constraints, which ensure the assumption of the rigidity of the body. Due to the assumption of the rigidity, the orthogonal rotation matrix $\boldsymbol{R} \in SO(3)$ satisfies the following conditions:

$$\boldsymbol{R}\boldsymbol{R}^T = \boldsymbol{I}, \quad \boldsymbol{R}^T = \boldsymbol{R}^{-1}, \quad \det(\boldsymbol{R}) = 1 \tag{2.33}$$

The rotational degrees of freedom of rigid body are calculated by $f = 9 - 6 = 3$.

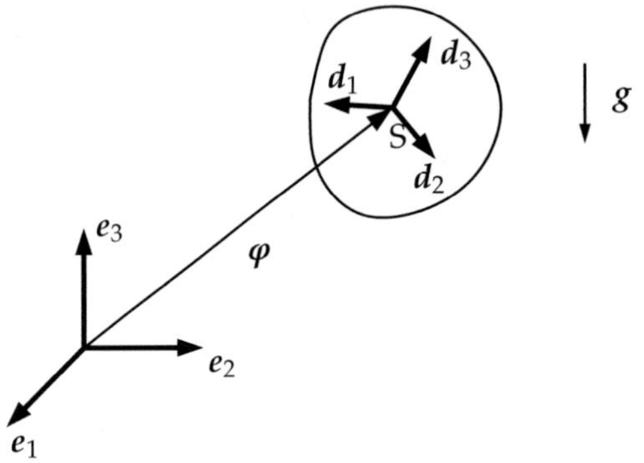

Figure 2.1: Spatial rigid body.

The configuration of the rigid body (Fig. 2.1) in three-dimensional Euclidean space can be described by the position vector of the body's center of mass $\boldsymbol{\varphi} \in \mathbb{R}^3$ and a right-handed body-fixed coordinate system $\{\boldsymbol{d}_i\}$, $\boldsymbol{d}_i \in \mathbb{R}^3$, $(i = 1, 2, 3)$. The vectors $\boldsymbol{d}_i$ are called the directors of the body and for simplicity the axes of the body-fixed director frame $\{\boldsymbol{d}_i\}$ are assumed to be aligned with the principal axes of the rigid body. Then the set of 12 redundant coordinates comprises the configuration vector

$$q = \begin{bmatrix} \boldsymbol{\varphi} \\ \boldsymbol{d}_1 \\ \boldsymbol{d}_2 \\ \boldsymbol{d}_3 \end{bmatrix} \tag{2.34}$$

which describes the configuration of the rigid body in the three-dimensional space. The configuration vector is specified relative to the inertial Cartesian basis $\{\boldsymbol{e}_i\}$. If a material point $\boldsymbol{X} = X_i \boldsymbol{e}_i$[1] belongs to the reference configuration $\mathcal{B}_0 \subset \mathbb{R}^3$ of the rigid

[1] The Einstein summation convention is used in the context.

body, the current spatial positon of $\boldsymbol{X} \in B_0$ at time t relative to the inertial Cartesian basis $\{\boldsymbol{e}_i\}$ is expressed by

$$\boldsymbol{x}(\boldsymbol{X}, t) = \boldsymbol{\varphi}(t) + X_i \boldsymbol{d}_i(t) \tag{2.35}$$

and the time derivative of $\boldsymbol{x}(\boldsymbol{X}, t)$ yields the velocity of the material point

$$\boldsymbol{v}(\boldsymbol{X}, t) = \dot{\boldsymbol{\varphi}}(t) + X_i \dot{\boldsymbol{d}}_i(t) \tag{2.36}$$

Thus the kinetic energy can be written as

$$T = \frac{1}{2} \int_{B_0} \boldsymbol{v}(\boldsymbol{X}, t) \cdot \boldsymbol{v}(\boldsymbol{X}, t) \rho(\boldsymbol{X}) dV \tag{2.37}$$

Since the axes of the body-fixed frame are aligned with the principal axes of the rigid body, the kinetic energy of the rigid body can be given by

$$T = \frac{1}{2} M_{\varphi} \dot{\boldsymbol{\varphi}} \cdot \dot{\boldsymbol{\varphi}} + \frac{1}{2} \sum_{i=1}^{3} E_i \dot{\boldsymbol{d}}_i \cdot \dot{\boldsymbol{d}}_i \tag{2.38}$$

where $\dot{\boldsymbol{\varphi}}$ and $\dot{\boldsymbol{d}}_i$ are redundant velocities, M_{φ} and E_i are expressed by

$$M_{\varphi} = \int_{B_0} \rho(\boldsymbol{X}) dV \tag{2.39a}$$

$$E_i = \int_{B_0} (X_i)^2 \rho(\boldsymbol{X}) dV \tag{2.39b}$$

The mass density at $\boldsymbol{X} \in B_0$ is denoted by $\rho(\boldsymbol{X})$, the total mass of the body is M_{φ} and the principal values of the Euler tensor with respect to the center of mass is expressed by E_i. The current Euler tensor with respect to the center of mass is given by

$$\boldsymbol{E} = \sum_{i=1}^{3} E_i \boldsymbol{d}_i \otimes \boldsymbol{d}_i \tag{2.40}$$

The Euler tensor is symmetric, positive-definite and can be related to the inertia tensor via the relationship

$$\boldsymbol{J} = (\mathrm{tr}\boldsymbol{E})\boldsymbol{I} - \boldsymbol{E} \tag{2.41}$$

According to Equations (2.18) and (2.38), the mass matix for a rigid body is given by

$$\boldsymbol{M} = \begin{bmatrix} M_{\varphi}\boldsymbol{I} & \mathbf{0} & \mathbf{0} & \mathbf{0} \\ \mathbf{0} & E_1\boldsymbol{I} & \mathbf{0} & \mathbf{0} \\ \mathbf{0} & \mathbf{0} & E_2\boldsymbol{I} & \mathbf{0} \\ \mathbf{0} & \mathbf{0} & \mathbf{0} & E_3\boldsymbol{I} \end{bmatrix} \tag{2.42}$$

in which each identity matrix $\boldsymbol{I}$ and zero matrix $\mathbf{0}$ has the dimension three. The above constant and diagonal 12×12 mass matrix $\boldsymbol{M}$ exhibits the specific characteristic of the rotationless formulation, whereas the mass matrix $\boldsymbol{\mathcal{M}}$ derived in the generalized coordinates formulation is configuration dependent.

Then the potential energy of the rigid body is given by

$$V = M_{\varphi}\, g\, \boldsymbol{\varphi} \cdot \boldsymbol{e}_3 \tag{2.43}$$

when the gravity is considered. Due to the assumption of the rigidity, the body-fixed frame keeps orthonormal for all times $t \in \mathbb{R}_{+}$. Thus, there are six independent internal constraints with associated constraint functions:

$$\boldsymbol{\Phi}_{\text{int}}(\boldsymbol{q}) = \begin{bmatrix} \frac{1}{2}(\boldsymbol{d}_1 \cdot \boldsymbol{d}_1) - 1 \\ \frac{1}{2}(\boldsymbol{d}_2 \cdot \boldsymbol{d}_2) - 1 \\ \frac{1}{2}(\boldsymbol{d}_3 \cdot \boldsymbol{d}_3) - 1 \\ \boldsymbol{d}_1 \cdot \boldsymbol{d}_2 \\ \boldsymbol{d}_1 \cdot \boldsymbol{d}_3 \\ \boldsymbol{d}_2 \cdot \boldsymbol{d}_3 \end{bmatrix} \tag{2.44}$$

According to $\boldsymbol{G}_{\text{int}} = \nabla_{\boldsymbol{q}} \boldsymbol{\Phi}_{\text{int}}(\boldsymbol{q})$, the internal constraints give rise to the corresponding constraint Jacobian:

$$\boldsymbol{G}_{\text{int}}(\boldsymbol{q}) = \begin{bmatrix} \mathbf{0}^T & \boldsymbol{d}_1^T & \mathbf{0}^T & \mathbf{0}^T \\ \mathbf{0}^T & \mathbf{0}^T & \boldsymbol{d}_2^T & \mathbf{0}^T \\ \mathbf{0}^T & \mathbf{0}^T & \mathbf{0}^T & \boldsymbol{d}_3^T \\ \mathbf{0}^T & \boldsymbol{d}_2^T & \boldsymbol{d}_1^T & \mathbf{0}^T \\ \mathbf{0}^T & \boldsymbol{d}_3^T & \mathbf{0}^T & \boldsymbol{d}_1^T \\ \mathbf{0}^T & \mathbf{0}^T & \boldsymbol{d}_3^T & \boldsymbol{d}_2^T \end{bmatrix} \tag{2.45}$$

in which the vector $\mathbf{0}$ has the dimension three and the constraint Jacobian matrix has the dimension 6×12. Additional details about the rigid body formulation can be found in [13, 17, 84, 87]. Note that the present rotationless formulation has similarities to the natural coordinates formulation advocated by García de Jalón et al. [37, 38].

Reduced formulation of the rigid body

Next the reduced dynamic equations (2.29a)–(2.29c) will be considered in depth for the spatial rigid body. First, the redundant velocities $\boldsymbol{v} \in \mathbb{R}^{12}$ of the rigid body are expressed in terms of the twist [3], which is composed of the translational velocity $\dot{\boldsymbol{\varphi}} \in \mathbb{R}^3$ of the body's center of mass and the angular velocity $\boldsymbol{\omega} \in \mathbb{R}^3$ of the rigid body. That is

$$\boldsymbol{t} = \begin{bmatrix} \dot{\boldsymbol{\varphi}} \\ \boldsymbol{\omega} \end{bmatrix} \tag{2.46}$$

Then the director velocities $\dot{\boldsymbol{d}}_i \in \mathbb{R}^3$ are expressed in terms of the angular velocity of the rigid body through

$$\dot{\boldsymbol{d}}_i = \boldsymbol{\omega} \times \boldsymbol{d}_i = -\boldsymbol{d}_i \times \boldsymbol{\omega} = -\widehat{\boldsymbol{d}}_i \boldsymbol{\omega} \tag{2.47}$$

Here, $\widehat{\boldsymbol{d}}_i$ denotes the skew-symmetric 3×3 matrix with the corresponding axial vector $\boldsymbol{d}_i \in \mathbb{R}^3$, that is

$$\widehat{\boldsymbol{d}}_i = \begin{bmatrix} 0 & -d_{i_3} & d_{i_2} \\ d_{i_3} & 0 & -d_{i_1} \\ -d_{i_2} & d_{i1} & 0 \end{bmatrix} \tag{2.48}$$

provided that

$$\boldsymbol{d}_i = \begin{bmatrix} d_{i_1} \\ d_{i_2} \\ d_{i_3} \end{bmatrix} \tag{2.49}$$

In view of Equation (2.26), the components of the twist $\boldsymbol{t} \in \mathbb{R}^6$ play the role of independent pseudo velocities of rigid body. The tangent space $T_{\boldsymbol{q}}Q = \ker\left(\boldsymbol{G}_{\text{int}}(\boldsymbol{q})\right)$ is determined and admissible velocities $\boldsymbol{v} \in T_{\boldsymbol{q}}Q$ are specified through the relation

$$\boldsymbol{v} = \boldsymbol{P}_{\text{int}}(\boldsymbol{q})\boldsymbol{t} \tag{2.50}$$

where the 12×6 null space matrix $\boldsymbol{P}_{\text{int}}$ for the free rigid body is given by

$$\boldsymbol{P}_{\text{int}}(\boldsymbol{q}) = \begin{bmatrix} \boldsymbol{I} & \boldsymbol{0} \\ \boldsymbol{0} & -\widehat{\boldsymbol{d}}_1 \\ \boldsymbol{0} & -\widehat{\boldsymbol{d}}_2 \\ \boldsymbol{0} & -\widehat{\boldsymbol{d}}_3 \end{bmatrix} \tag{2.51}$$

Here, each $\mathbf{0}$ and $\boldsymbol{I}$ matrix has the dimesion three. In view of Equation (2.45), the condition $\boldsymbol{G}_{\text{int}}\boldsymbol{P}_{\text{int}} = \mathbf{0}$ is fully satisfied.

Then the reduced mass matrix $\widetilde{\boldsymbol{M}}$ for the rigid body can be calculated through

$$\widetilde{\boldsymbol{M}} = \boldsymbol{P}_{\text{int}}^T \boldsymbol{M} \boldsymbol{P}_{\text{int}} = \begin{bmatrix} M_\varphi \boldsymbol{I} & \mathbf{0} \\ \mathbf{0} & -\sum_{i=1}^3 E_i (\widehat{\boldsymbol{d}}_i)^2 \end{bmatrix} \tag{2.52}$$

Using the identity $(\widehat{\boldsymbol{d}}_i)^2 = \boldsymbol{d}_i \otimes \boldsymbol{d}_i - (\boldsymbol{d}_i \cdot \boldsymbol{d}_i)\boldsymbol{I}$, Equation (2.41) leads to

$$-\sum_{i=1}^{3} E_i (\widehat{\boldsymbol{d}}_i)^2 = -\sum_{i=1}^{3} E_i \left[\boldsymbol{d}_i \otimes \boldsymbol{d}_i - (\boldsymbol{d}_i \cdot \boldsymbol{d}_i)\boldsymbol{I}\right] = -\boldsymbol{E} + (\text{tr}\boldsymbol{E})\boldsymbol{I} = \boldsymbol{J} \tag{2.53}$$

with the customary inertia tensor of the rigid body $\boldsymbol{J}$. Thus, the reduced mass matrix can also be written as

$$\widetilde{\boldsymbol{M}} = \begin{bmatrix} M_\varphi \boldsymbol{I} & \mathbf{0} \\ \mathbf{0} & \boldsymbol{J} \end{bmatrix} \tag{2.54}$$

In addition, $\boldsymbol{P}^T \boldsymbol{M} \dot{\boldsymbol{P}} \boldsymbol{t}$ in Equation (2.29b) is expressed as

$$\boldsymbol{P}^T \boldsymbol{M} \dot{\boldsymbol{P}} \boldsymbol{t} = \begin{bmatrix} \mathbf{0} \\ -\boldsymbol{\omega} \times \left(\sum_{i=1}^3 E_i \boldsymbol{d}_i \otimes \boldsymbol{d}_i\right) \boldsymbol{\omega} \end{bmatrix} = \begin{bmatrix} \mathbf{0} \\ \boldsymbol{\omega} \times \boldsymbol{J}\boldsymbol{\omega} \end{bmatrix} \tag{2.55}$$

by using Equation (2.41). The last term in Equation (2.29b) is calculated by

$$\boldsymbol{P}^T \nabla V(\boldsymbol{q}) = \begin{bmatrix} \frac{\partial V}{\partial \boldsymbol{\varphi}} \\ \sum_{i=1}^3 \left(\boldsymbol{d}_i \times \frac{\partial V}{\partial \boldsymbol{d}_i}\right) \end{bmatrix} = -\begin{bmatrix} \overline{\boldsymbol{f}} \\ \overline{\boldsymbol{m}} \end{bmatrix} \tag{2.56}$$

where $\overline{\boldsymbol{f}}$ and $\overline{\boldsymbol{m}}$ are the resultant force and torque relative to the rigid body's center of mass. In the end the reduced equations of motion (2.29a)–(2.29c) for the free rigid body are written as

$$M_\varphi \ddot{\boldsymbol{\varphi}} = \overline{\boldsymbol{f}} \tag{2.57a}$$

$$\boldsymbol{J}\dot{\boldsymbol{\omega}} + \boldsymbol{\omega} \times \boldsymbol{J}\boldsymbol{\omega} = \overline{\boldsymbol{m}} \tag{2.57b}$$

which are the Newton-Euler equations for rigid bodies.

Generalized coordinates, such as Euler angles or the rotation vector $\boldsymbol{\theta} \in \mathbb{R}^3$, can be employed to describe the orientation of the rigid body instead of the directors $\boldsymbol{d}_i$. Therefore, the number of redundant coordinates $\boldsymbol{q} \in \mathbb{R}^{12}$ can be further reduced by reparameterization of the configuration space. A rotation matrix $\boldsymbol{R} \in SO(3)$ is introduced, which is parameterized in terms of $\boldsymbol{\theta} \in \mathbb{R}^3$, and the directors of the rigid body can be expressed by

$$\boldsymbol{d}_i = \boldsymbol{R}(\boldsymbol{\theta})\boldsymbol{d}_{i_0} \tag{2.58}$$

where $\boldsymbol{d}_{i_0}$ denotes the initial directors relative to the inertial Cartesian basis $\{\boldsymbol{e}_i\}$ at time $t = 0$. The rotation matrix can be calculated through the Rodrigues formula [42, 72], that may be expressed by the exponential map representation:

$$\boldsymbol{R}(\boldsymbol{\theta}) = \exp(\widehat{\boldsymbol{\theta}}) = \boldsymbol{I} + \frac{\sin\|\boldsymbol{\theta}\|}{\|\boldsymbol{\theta}\|}\widehat{\boldsymbol{\theta}} + \frac{1-\cos\|\boldsymbol{\theta}\|}{\|\boldsymbol{\theta}\|^2}\widehat{\boldsymbol{\theta}}^2 \tag{2.59}$$

After the reparametrization of the unknowns the configuration of the free rigid body is specified by six coordinates $\boldsymbol{\mu} = (\boldsymbol{\varphi}, \boldsymbol{\theta}) \in U \subset \mathbb{R}^3 \times \mathbb{R}^3$. A mapping $\boldsymbol{F} : U \mapsto Q$ is given by

$$\boldsymbol{q} = \boldsymbol{F}(\boldsymbol{\mu}) = \begin{bmatrix} \boldsymbol{\varphi} \\ \exp(\widehat{\boldsymbol{\theta}})\boldsymbol{d}_{1_0} \\ \exp(\widehat{\boldsymbol{\theta}})\boldsymbol{d}_{2_0} \\ \exp(\widehat{\boldsymbol{\theta}})\boldsymbol{d}_{3_0} \end{bmatrix} \tag{2.60}$$

2.3.4 Kinematic pairs

Rigid multibody systems consist of rigid bodies interconnected by different types of joints, which can be classified into two groups: lower pairs and higher pairs. The joints with surface contact are referred to as lower pairs and the joints with point or line contact are called higher pairs [42]. The basic kinematic pairs have been treated thoroughly in several works [13, 79, 84, 87]. Here only the revolute pair and the prismatic pair will be introduced so that they can be used in the three-dimensional rotary crane example in the sequel.

A simple multibody system is now considered, which consists of two rigid bodies coupled by lower kinematic pairs. The configuration of the α-th rigid body[2] can be characterized by redundant coordinates $\boldsymbol{q}^\alpha \in \mathbb{R}^{12}$ $(\alpha = 1, 2)$, which has been given in Equation (2.34). Then the configuration of two rigid bodies can be characterized by 24 redundant coordinates. Thus, the configuration vector is written as

[2] In the framework of the rotationless formulation the superscript denotes the respective rigid body.

$$q = \begin{bmatrix} q^1 \\ q^2 \end{bmatrix} \tag{2.61}$$

Due to the assumption of the rigidity, each rigid body is subject to six internal constraints. For the two-body multibody system, 12 internal constraint functions are described by

$$\boldsymbol{\Phi}_{\text{int}}(q) = \begin{bmatrix} \boldsymbol{\Phi}^1_{\text{int}}(q^1) \\ \boldsymbol{\Phi}^2_{\text{int}}(q^2) \end{bmatrix} \tag{2.62}$$

and the associated constraint Jacobian $\boldsymbol{G}_{\text{int}}(q) \in \mathbb{R}^{12\times 24}$ is given by

$$\boldsymbol{G}_{\text{int}}(q) = \begin{bmatrix} \boldsymbol{G}^1_{\text{int}}(q^1) & \mathbf{0} \\ \mathbf{0} & \boldsymbol{G}^2_{\text{int}}(q^2) \end{bmatrix} \tag{2.63}$$

Here, $\boldsymbol{\Phi}^\alpha_{\text{int}}(q^\alpha)$ and $\boldsymbol{G}^\alpha_{\text{int}}(q^\alpha)$ have already been given in Equations (2.44) and (2.45), respectively.

In addition, the connection of two rigid bodies by a specific joint leads to the external constraints expressed by constraint functions $\boldsymbol{\Phi}_{\text{ext}}(q)$. Accordingly, the kinematic pair can be characterized by the constraint functions

$$\boldsymbol{\Phi}(q) = \begin{bmatrix} \boldsymbol{\Phi}_{\text{int}}(q) \\ \boldsymbol{\Phi}_{\text{ext}}(q) \end{bmatrix} \tag{2.64}$$

and the corresponding constraint Jacobian

$$\boldsymbol{G}(q) = \begin{bmatrix} \boldsymbol{G}_{\text{int}}(q) \\ \boldsymbol{G}_{\text{ext}}(q) \end{bmatrix} \tag{2.65}$$

The equations of motion of the kinematic pair can then be expressed by Equation (2.20a)–(2.20c). The constant mass matrix $\boldsymbol{M} \in \mathbb{R}^{24\times 24}$ then reads

$$\boldsymbol{M} = \begin{bmatrix} \boldsymbol{M}^1 & \mathbf{0} \\ \mathbf{0} & \boldsymbol{M}^2 \end{bmatrix} \tag{2.66}$$

in which the submatrix $\boldsymbol{M}^\alpha \in \mathbb{R}^{12\times 12}$ is given in Equation (2.42).

Next two primitive kinematic pairs will be presented in the framework of the rotationless formulation.

Revolute pair

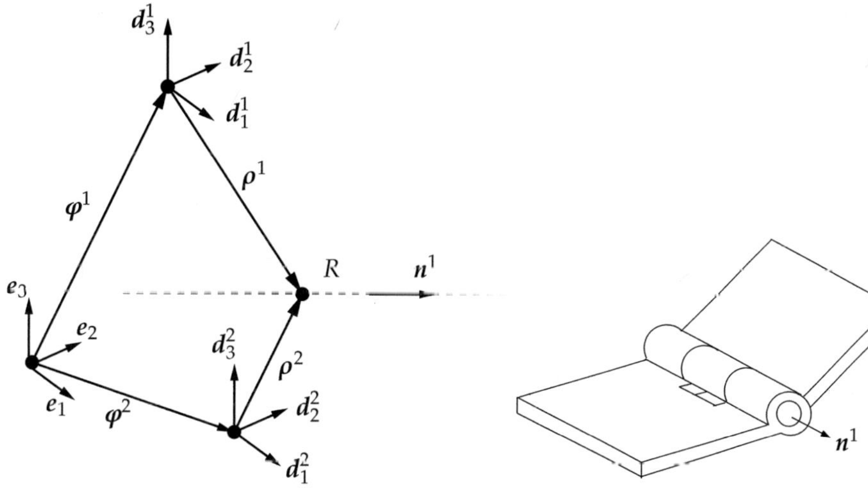

Figure 2.2: Revolute pair.

The revolute pair is a basic kinematic joint for multibody systems. As shown in Fig. 2.2, it consists of two rigid bodies ($\alpha = 1, 2$) and the corresponding director frame $\{\boldsymbol{d}_i^\alpha\}$ ($i = 1, 2, 3$) is fixed to each individual body. The location of the joint is specified by coordinates ρ_i^α with respect to the individual body frame $\{\boldsymbol{d}_i^\alpha\}$. That is

$$\boldsymbol{\rho}^\alpha = \rho_i^\alpha \boldsymbol{d}_i^\alpha \tag{2.67}$$

Then a unit vector $\boldsymbol{n}^1$ is introduced, which is fixed to the first body and specified by coordinates n_i^1 with respect to the director frame $\left\{\boldsymbol{d}_i^1\right\}$:

$$\boldsymbol{n}^1 = n_i^1 \boldsymbol{d}_i^1 \tag{2.68}$$

Additionally, two more vectors (see Fig. 2.3) can be introduced and defined as

$$\boldsymbol{m}_\alpha^1 = (m_\alpha^1)_i \boldsymbol{d}_i^1 \tag{2.69}$$

such that $\left\{\boldsymbol{m}_1^1, \boldsymbol{m}_2^1, \boldsymbol{n}^1\right\}$ constitute a right-handed orthonormal frame. For the revolute pair, the axis of rotation of the second body relative to the first body is specified by the unit vector $\boldsymbol{n}^1$ and the relative degree of freedom is one. The revolute pair entails 5 external constraint functions, which can be written in the following form

$$\boldsymbol{\Phi}_{\text{ext}}(\boldsymbol{q}) = \begin{bmatrix} \boldsymbol{\varphi}^2 + \boldsymbol{\rho}^2 - \boldsymbol{\varphi}^1 - \boldsymbol{\rho}^1 \\ \boldsymbol{n}^1 \cdot \boldsymbol{d}_1^2 - \eta_1 \\ \boldsymbol{n}^1 \cdot \boldsymbol{d}_2^2 - \eta_2 \end{bmatrix} \tag{2.70}$$

where η_1 and η_2 are constant and need to be consistent with the initial conditions. Moreover, the corresponding constraint Jacobian of the external constraint functions is given by the 5×24 matrix

$$\boldsymbol{G}_{\text{ext}}(\boldsymbol{q}) = \begin{bmatrix} -\boldsymbol{I} & -\rho_1^1\boldsymbol{I} & -\rho_2^1\boldsymbol{I} & -\rho_3^1\boldsymbol{I} & \boldsymbol{I} & \rho_1^2\boldsymbol{I} & \rho_2^2\boldsymbol{I} & \rho_3^2\boldsymbol{I} \\ \boldsymbol{0}^T & n_1^1(\boldsymbol{d}_1^2)^T & n_2^1(\boldsymbol{d}_1^2)^T & n_3^1(\boldsymbol{d}_1^2)^T & \boldsymbol{0}^T & (\boldsymbol{n}^1)^T & \boldsymbol{0}^T & \boldsymbol{0}^T \\ \boldsymbol{0}^T & n_1^1(\boldsymbol{d}_2^2)^T & n_2^1(\boldsymbol{d}_2^2)^T & n_3^1(\boldsymbol{d}_2^2)^T & \boldsymbol{0}^T & \boldsymbol{0}^T & (\boldsymbol{n}^1)^T & \boldsymbol{0}^T \end{bmatrix} \tag{2.71}$$

where the vector $\boldsymbol{0}$ and the identity matrix $\boldsymbol{I}$ have the dimension three. In conclusion, for the two-body multibody system with revolute pair, there are 24 redundant coordinates expressed in Equation (2.61), which are subject to 12 internal constraint functions (2.62) and 5 external constraint functions (2.70). The degrees of freedom of the revolute pair is then calculated by $f = 24 - 12 - 5 = 7$.

Prismatic pair

The prismatic pair as shown in Fig. 2.3, is also used as a basic kinematic joint to connect rigid bodies. In the two-body multibody system with prismatic pair, the translational motion of the second body relative to the first body occurs along the translational axis specified by the unit vector $\boldsymbol{n}^1$, which is fixed at the first body and can be expressed by Equation (2.68). Therefore, the prismatic joint has one relative degree of freedom.

Analogously, the prismatic pair entails 5 external constraint functions, which may be written in the form

$$\boldsymbol{\Phi}_{\text{ext}}(\boldsymbol{q}) = \begin{bmatrix} \boldsymbol{m}_1^1 \cdot (\boldsymbol{\varphi}^2 + \boldsymbol{\rho}^2 - \boldsymbol{\varphi}^1 - \boldsymbol{\rho}^1) \\ \boldsymbol{m}_2^1 \cdot (\boldsymbol{\varphi}^2 + \boldsymbol{\rho}^2 - \boldsymbol{\varphi}^1 - \boldsymbol{\rho}^1) \\ \boldsymbol{d}_1^1 \cdot \boldsymbol{d}_2^2 - \eta_1 \\ \boldsymbol{d}_2^1 \cdot \boldsymbol{d}_3^2 - \eta_2 \\ \boldsymbol{d}_3^1 \cdot \boldsymbol{d}_1^2 - \eta_3 \end{bmatrix} \tag{2.72}$$

where η_1, η_2 and η_3 are constant and need to be consistent with the initial conditions.

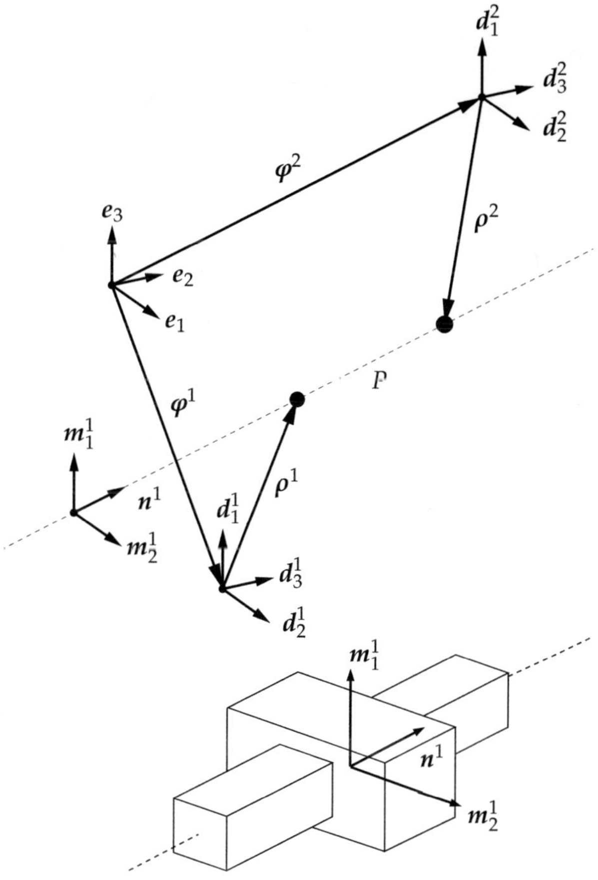

Figure 2.3: Prismatic pair.

Similarly, the prismatic pair has 7 degrees of freedom. Furthermore, the corresponding constraint Jacobian of the external constraint functions is given by the following 5×24 matrix:

$$G_{\text{ext}}(q) = \begin{bmatrix} -(m_1^1)^T & G_{11}^T & G_{12}^T & G_{13}^T & (m_1^1)^T & \rho_1^2(m_1^1)^T & \rho_2^2(m_1^1)^T & \rho_3^2(m_1^1)^T \\ -(m_2^1)^T & G_{21}^T & G_{22}^T & G_{23}^T & (m_2^1)^T & \rho_1^2(m_2^1)^T & \rho_2^2(m_2^1)^T & \rho_3^2(m_2^1)^T \\ 0^T & (d_2^2)^T & 0^T & 0^T & 0^T & 0^T & (d_1^1)^T & 0^T \\ 0^T & 0^T & (d_3^2)^T & 0^T & 0^T & 0^T & 0^T & (d_2^1)^T \\ 0^T & 0^T & 0^T & (d_1^2)^T & 0^T & (d_3^1)^T & 0^T & 0^T \end{bmatrix} \tag{2.73}$$

with the expression

$$G_{\alpha i} = (m_\alpha^1)_i\,(\varphi^2 + \rho^2 - \varphi^1 - \rho^1) - \rho_i^1 m_\alpha^1 \tag{2.74}$$

for $\alpha = 1,2$ and $i = 1,2,3$. More details about the null space reduction and reparametrization of unknowns can be found in [13].

2.3.5 Coordinate augmentation

The rotationless formulation circumvents the use of rotational parameters. However, in many practical applications the rotational degrees of freedom with associated torques need to be considered in the formulation of multibody dynamics. A specific coordinate augmentation technique [19, 88, 87] can be applied to incorporate the rotational parameters with associated torques into the rotationless formulation of multibody systems. It is to be noted that the coordinate augmentation does not destroy the advantageous features of the rotationless formulation.

In the following section a three-dimensional rotary crane example will demonstrate the application of the coordinate augmentation technique.

2.4 Three-dimensional rotary crane

There are many different types of cranes like overhead cranes or rotary cranes, which are widely used in various fields like in the transportation or construction industries. In practical applications, they are operated by the human crane operator to move a payload from the initial position to the end position along a trajectory in the working space. The operation needs to avoid the obstacles and sways of the payload [28, 59] and requires the motion planning for the payload position.

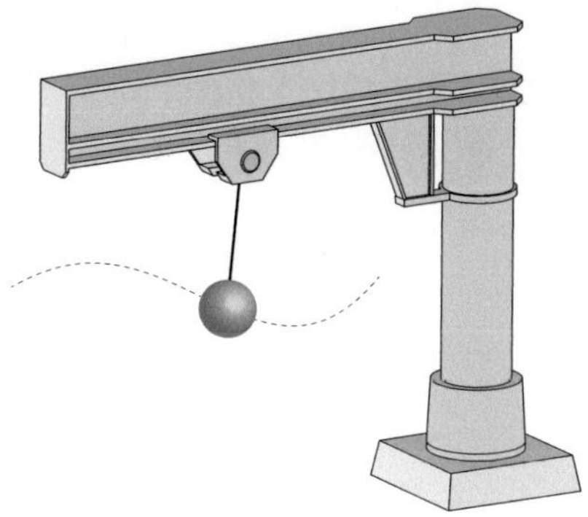

Figure 2.4: The three-dimensional rotary crane example.

In the inverse dynamics simulation, cranes are underactuated ($a < f$) mechanical systems, in which the number a of control inputs/outputs is smaller than the number f of degrees of freedom [26]. Besides, there are also fully actuated ($a = f$) or overactuated ($a > f$) mechanical systems. The subsequent chapters will focus on the inverse dynamics simulation of underactuated systems.

The three-dimensional rotary crane example depicted in Fig. 2.4 has originally been dealt with in the inverse dynamics simulation by Blajer and Kołodziejczyk [28], where generalized coordinates were used to formulate the dynamics of the system. As described in the previous sections, for the same rotary crane example, either the generalized coordinates formulation in terms of minimal coordinates or the rotationless formulation in terms of redundant coordinates can be applied to perform the forward dynamics simulations [93]. Both formulations will be provided in the following.

2.4.1 Generalized coordinates formulation

The model of the rotary crane, as shown in Fig. 2.5, is considered as a rigid multibody system, that is comprised of three rigid bodies and one payload. The payload is here assumed as a point mass (mass m_4). The first rigid body consists of the girder bridge and the pillar, and its moment of inertia relative to the rotation axis $\boldsymbol{d}_3^1$[3] is expressed by J_3^1. Here the vector $\boldsymbol{d}_3^1$ is identical to the unit vector e_3 in the direction of Z-axis (see also Fig. 2.9). The second rigid body is the trolley (mass m_2), which undergoes

[3] The subscript indicates which axis of the body frame is regarded as the rotation axis of the rigid body, and the superscript indicates which rigid body is considered.

the translational motion on the girder bridge. The third rigid body is the winch (mass m_3), which is contained in the trolley. The moment of inertia of the winch relative to the rotation axis $\boldsymbol{d}_2^3$ (see also Fig. 2.9) is given by J_2^3.

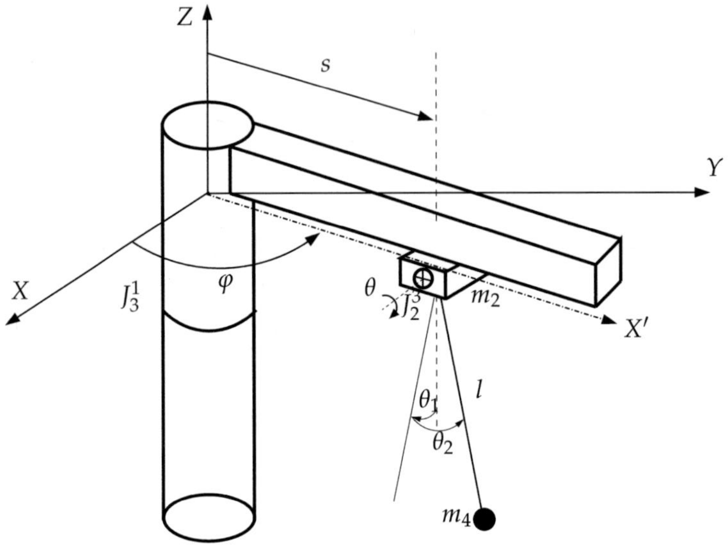

Figure 2.5: The rotary crane model in terms of 5 generalized coordinates.

The rotary crane model is an underactuated mechanical system with 5 degrees of freedom, that is, $f = 5$. Then the configuration of the system can be described by a set of generalized coordinates:

$$\boldsymbol{\mu} = \begin{bmatrix} \varphi & s & l & \theta_1 & \theta_2 \end{bmatrix}^T \tag{2.75}$$

Here, the rotation angle of the girder bridge about the Z-axis relative to the X-axis is given by φ, the position of the trolley on the girder bridge is specified by s, the length of the hoisting cable is denoted by l which connects the payload with the winch, and the swing angles depicted in Fig. 2.5 are given by θ_1 and θ_2. Moreover, the rotation angle θ of the winch, as shown in Fig. 2.5, is related to the cable length l and can be expressed by

$$\theta = \frac{l - l_0}{r_w} \tag{2.76}$$

in which the winch radius is denoted by r_w and the initial cable length is l_0.

Position vectors and finite rotations

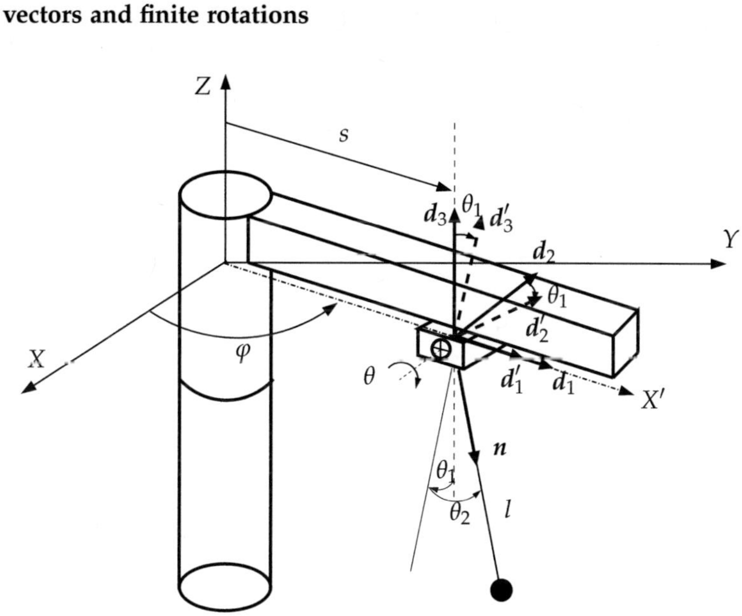

Figure 2.6: The rotary crane model with body frames $\{d_1, d_2, d_3\}$ and $\{d'_1, d'_2, d'_3\}$.

During the movement of the rotary crane the first body only has the rotational motion about the Z-axis (see Fig. 2.6). The center of mass of the first body is assumed to be at the origin of the inertial Cartesian coordinate system. Then the position vector of the first body reads

$$\boldsymbol{\varphi}^1 = \begin{bmatrix} 0 \\ 0 \\ 0 \end{bmatrix} \tag{2.77}$$

The position vectors of the trolley and the winch read

$$\boldsymbol{\varphi}^2 = \boldsymbol{\varphi}^3 = \begin{bmatrix} s\cos\varphi \\ s\sin\varphi \\ 0 \end{bmatrix} \tag{2.78}$$

The position vector of the payload has been provided by Blajer and Kołodziejczyk [28] and it reads

$$\boldsymbol{\varphi}^4 = \begin{bmatrix} (s + l\sin\theta_2)\cos\varphi + l\cos\theta_2\sin\theta_1\sin\varphi \\ (s + l\sin\theta_2)\sin\varphi - l\cos\theta_2\sin\theta_1\cos\varphi \\ -l\cos\theta_2\cos\theta_1 \end{bmatrix} \tag{2.79}$$

The position vector of the payload can also be calculated by

$$\boldsymbol{\varphi}^4 = \boldsymbol{\varphi}^3 + l\boldsymbol{n} \tag{2.80}$$

Here, a unit vector (see Fig. 2.6) is given by $\boldsymbol{n}$, which is directed along the cable and points to the payload .

A sequence of finite rotations about the body-fixed frame axis can be applied to calculate the unit vector $\widetilde{\boldsymbol{n}}$:

$$\widetilde{\boldsymbol{n}} = \boldsymbol{R}(-\boldsymbol{e}_3) = \boldsymbol{R}(\boldsymbol{\theta}_1)\boldsymbol{R}(\boldsymbol{\theta}_2)(-\boldsymbol{e}_3) = \exp(\widehat{\boldsymbol{\theta}}_1)\exp(\widehat{\boldsymbol{\theta}}_2)(-\boldsymbol{e}_3) = \begin{bmatrix} \sin\theta_2 \\ -\sin\theta_1\cos\theta_2 \\ -\cos\theta_1\cos\theta_2 \end{bmatrix} \tag{2.81}$$

where the components of $\widetilde{\boldsymbol{n}}$ are expressed in the initial body frame $\{\boldsymbol{d}_1, \boldsymbol{d}_2, \boldsymbol{d}_3\}$ and $\boldsymbol{e}_3 = \begin{bmatrix} 0 & 0 & 1 \end{bmatrix}^T$.

For Equation (2.81) it is necessary to first calculate the rotation matrix. At first, the body frame of the winch is rotated clockwise from its initial orientation $\{\boldsymbol{d}_1, \boldsymbol{d}_2, \boldsymbol{d}_3\}$ to the current orientation $\{\boldsymbol{d}_1', \boldsymbol{d}_2', \boldsymbol{d}_3'\}$ by an angle $(-\theta_1)$ about the body-fixed frame axis $\boldsymbol{d}_1$ of the winch. The rotation matrix of the first rotation is then expressed by

$$\boldsymbol{R}(\boldsymbol{\theta}_1) = \exp(\widehat{\boldsymbol{\theta}}_1) = \exp(\theta_1\widehat{\boldsymbol{d}}_1) = \begin{bmatrix} 1 & 0 & 0 \\ 0 & \cos\theta_1 & \sin\theta_1 \\ 0 & -\sin\theta_1 & \cos\theta_1 \end{bmatrix} \tag{2.82}$$

Then the body frame of the winch is rotated clockwise about the body-fixed frame axis $\boldsymbol{d}_2'$ by an angle of $(-\theta_2)$. The rotation matrix of the second rotation is given by

$$\boldsymbol{R}(\boldsymbol{\theta}_2) = \exp(\widehat{\boldsymbol{\theta}}_2) = \exp(\theta_2\widehat{\boldsymbol{d}_2'}) = \begin{bmatrix} \cos\theta_2 & 0 & -\sin\theta_2 \\ 0 & 1 & 0 \\ \sin\theta_2 & 0 & \cos\theta_2 \end{bmatrix} \tag{2.83}$$

Next the unit vector $\boldsymbol{n}$ expressed in the inertial Cartesian coordinate system is calculated by

$$\boldsymbol{n} = \widetilde{n}_1\,\boldsymbol{d}_1 + \widetilde{n}_2\,\boldsymbol{d}_2 + \widetilde{n}_3\,\boldsymbol{d}_3 = (\sin\theta_2)\,\boldsymbol{d}_1 - (\sin\theta_1\cos\theta_2)\,\boldsymbol{d}_2 - (\cos\theta_1\cos\theta_2)\,\boldsymbol{d}_3 \tag{2.84}$$

where the unit vectors of the initial body frame $\{\boldsymbol{d}_1, \boldsymbol{d}_2, \boldsymbol{d}_3\}$ are expressed by

$$d_1 = \begin{bmatrix} \cos\varphi \\ \sin\varphi \\ 0 \end{bmatrix} \qquad d_2 = \begin{bmatrix} -\sin\varphi \\ \cos\varphi \\ 0 \end{bmatrix} \qquad d_3 = \begin{bmatrix} 0 \\ 0 \\ 1 \end{bmatrix} \tag{2.85}$$

At last, inserting the unit vector $\boldsymbol{n}$ into Equation (2.80) yields the same expression as provided in Equation (2.79).

The Lagrangian function

The total kinetic energy of the three rigid bodies and one mass point is given by

$$T = T_1 + T_2 + T_3 + T_4 \tag{2.86}$$

with the energy components

$$T_1 = \frac{1}{2} J_3^1 \dot{\varphi}^2 \tag{2.87a}$$

$$T_2 = \frac{1}{2} m_2 \dot{s}^2 + \frac{1}{2} (J_3^2 + m_2 s^2) \dot{\varphi}^2 \tag{2.87b}$$

$$T_3 = \frac{1}{2} m_3 \dot{s}^2 + \frac{1}{2} (J_3^3 + m_3 s^2) \dot{\varphi}^2 + \frac{1}{2} J_2^3 \left(\frac{\dot{l}}{r_w}\right)^2 \tag{2.87c}$$

$$T_4 = \frac{1}{2} m_4 \dot{\boldsymbol{\varphi}}^4 \cdot \dot{\boldsymbol{\varphi}}^4 \tag{2.87d}$$

The potential energy of the rotary crane is given by

$$V = m_4 g \boldsymbol{\varphi}^4 \cdot \boldsymbol{e}_3 \tag{2.88}$$

Then the Lagrangian of the mechanical system is calculated by Equation (2.2) and the Lagrange's equations of the second kind are applied to obtain the ODEs (2.17a)–(2.17b) for the rotary crane model. It is to be noted that symbolic manipulations are used to compute the differentiation of the Lagrangian.

Numerical discretization

To solve the ordinary differential equations of the rotary crane model, the second order accurate mid-point-type rule is used and its application yields the discretized version of (2.17a)–(2.17b)

$$\mu_{n+1} - \mu_n = \Delta t \, \nu_{n+\frac{1}{2}} \tag{2.89a}$$

$$\mathcal{M}(\mu_{n+\frac{1}{2}}) \, (\nu_{n+1} - \nu_n) = \Delta t \, f(\mu_{n+\frac{1}{2}}, \nu_{n+\frac{1}{2}}) \tag{2.89b}$$

with the time step size $\Delta t = t_{n+1} - t_n$ in a representative time interval $[t_n, t_{n+1}]$. It should be noted that $(\bullet)_n$ and $(\bullet)_{n+1}$ represent the evaluation of the corresponding vector at the time point t_n and t_{n+1}, respectively. Moreover, $(\bullet)_{n+\frac{1}{2}} = \frac{1}{2}\left[(\bullet)_n + (\bullet)_{n+1}\right]$ denotes the mid-point evaluation of the corresponding vector.

Forward dynamics simulation

In the forward dynamics simulation, the initial configuration of the rotary crane is specified by the generalized coordinates

$$\mu_0 = \begin{bmatrix} 0 & 1.5\,\text{m} & 5\,\text{m} & 0 & 0 \end{bmatrix}^T \tag{2.90}$$

and the initial generalized velocities are given by

$$\nu_0 = \begin{bmatrix} 1\,\text{m/s} & 0 & 0 & 0 & 0 \end{bmatrix}^T \tag{2.91}$$

During the simulation time of $t = 1\,\text{s}$, for example, the girder bridge rotates with an initial angular velocity $\dot{\varphi} = 1\,\text{m/s}$. As there are no other external forces and torques acting on the mechanical system except the gravitational force, the payload falls down. Table 2.1 presents the data of mass and moment of inertia for each body of the rotary crane. In the numerical experiment, the winch radius is $r_w = 0.1\,\text{m}$, the gravitational acceleration is $g = 9.81\,\text{m/s}^2$ and the masses and moments of inertia are: $m_2 = 50\,\text{kg}$, $m_3 = 3\,\text{kg}$, $m_4 = 10\,\text{kg}$, $J_3^1 = 16.67\,\text{kg}\cdot\text{m}^2$, $J_3^2 = 2.08\,\text{kg}\cdot\text{m}^2$, $J_3^3 = 0.26\,\text{kg}\cdot\text{m}^2$, $J_2^3 = 0.02\,\text{kg}\cdot\text{m}^2$.

body	m [kg]	J_1 [kg · m^2]	J_2 [kg · m^2]	J_3 [kg · m^2]
1	100	216.67	216.67	16.67
2	50	2.08	2.08	2.08
3	3	0.26	0.02	0.26
4	10	—	—	—

Table 2.1: Data of mass and moment of inertia for each body of the rotary crane.

Numerical results

The numerical results of the forward dynamics simulation of the rotary crane are presented in Fig. 2.7. They show that the total energy of the system and the Z-axis component of the angular momentum L_z are conserved quantities because this rotary crane model is a conservative system with rotational symmetry about the Z-axis. Some snapshots of the movement of the rotary crane are presented in Fig. 2.8. It is to be noted that in the simulation the origin of the inertial Cartesian coordinate system XYZ is placed at the half height position of the pillar rather than at the top of it.

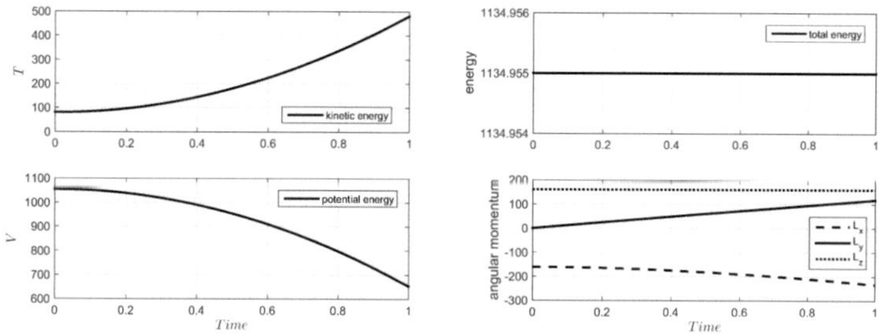

Figure 2.7: Generalized coordinates formulation: Energy and angular momentum of the rotary crane with the time step of $\Delta t = 10^{-4}$ s.

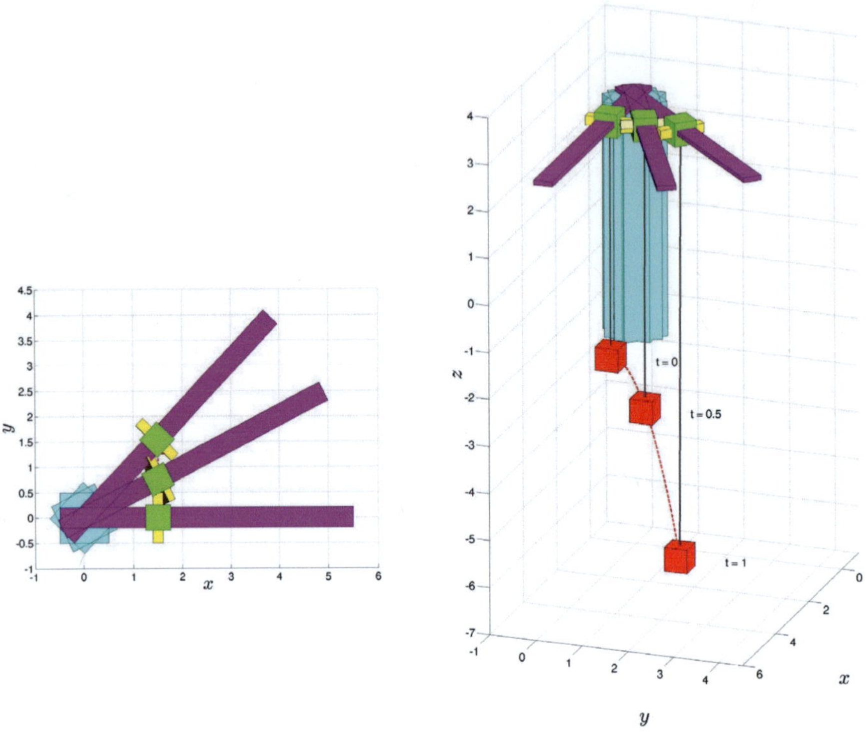

Figure 2.8: Generalized coordinates formulation: Snapshots of the simulation of the rotary crane at specific time points.

2.4.2 Rotationless formulation

The rotationless formulation in terms of redundant coordinates will be applied to formulate the numerical model of the rotary crane, which is shown in Fig. 2.9.

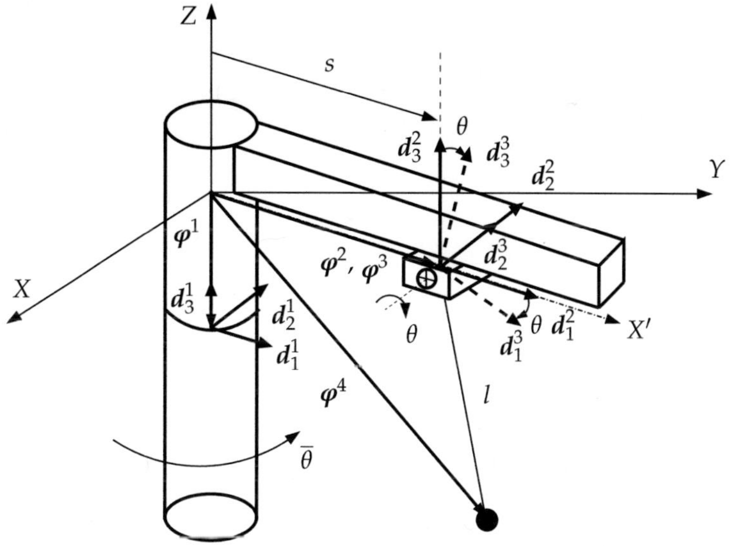

Figure 2.9: The model of the rotary crane with the position vectors φ^I, φ^4 and the body frames $\left\{d_1^I, d_2^I, d_3^I\right\}$, $I = 1,2,3$.

Rigid bodies

Proceeding along the lines of Uhlar and Betsch [88], the configuration of each rigid body ($I = 1,2,3$) and the point mass can be specified by the configuration vector $q^I \in \mathbb{R}^{12}$ in Equation (2.34) and $q^4 \in \mathbb{R}^3$. Then the configuration of the rotary crane model can be described by 39 redundant coordinates, which comprise the configuration vector

$$\bar{q} = \begin{bmatrix} q^1 \\ q^2 \\ q^3 \\ q^4 \end{bmatrix} \tag{2.92}$$

with each component

$$q^I = \begin{bmatrix} \varphi^I \\ d_1^I \\ d_2^I \\ d_3^I \end{bmatrix} \qquad (I = 1,2,3) \tag{2.93}$$

and the position vector of load mass

$$q^4 = \varphi^4 \tag{2.94}$$

Constraints

Due to the assumption of rigidity, each rigid body gives rise to six internal constraints (Equation (2.44)). Three rigid bodies lead to 18 internal constraints. Besides, the external constraints describing the joints should also be considered in the model. Accordingly, the first body is connected to the ground via a revolute joint. The external constraint functions are written as

$$\boldsymbol{\Phi}_{\text{rev1}}(\overline{q}) = \begin{bmatrix} \varphi^1 \\ d_1^1 \cdot e_3 \\ d_2^1 \cdot e_3 \end{bmatrix} \tag{2.95}$$

The associated constraint Jacobian is given by

$$G_{\text{rev1}}(\overline{q}) = \begin{bmatrix} I & 0_{3\times 3} & 0_{3\times 3} & 0_{3\times 3} & 0_{3\times 12} & 0_{3\times 12} & 0_{3\times 3} \\ 0^T & (e_3)^T & 0^T & 0^T & 0_{1\times 12} & 0_{1\times 12} & 0_{1\times 3} \\ 0^T & 0^T & (e_3)^T & 0^T & 0_{1\times 12} & 0_{1\times 12} & 0_{1\times 3} \end{bmatrix} \tag{2.96}$$

The trolley is then connected to the girder bridge by a prismatic joint. The external constraint functions are written as

$$\boldsymbol{\Phi}_{\text{pris}}(\overline{q}) = \begin{bmatrix} d_2^1 \cdot \varphi^2 \\ d_3^1 \cdot \varphi^2 \\ d_1^1 \cdot d_2^2 \\ d_2^1 \cdot d_3^2 \\ d_3^1 \cdot d_1^2 \end{bmatrix} \tag{2.97}$$

The corresponding constraint Jacobian is given by

$$G_{\text{pris}}(\overline{q}) = \begin{bmatrix} 0^T & 0^T & (\varphi^2)^T & 0^T & (d_2^1)^T & 0^T & 0^T & 0^T & 0_{1\times 12} & 0_{1\times 3} \\ 0^T & 0^T & 0^T & (\varphi^2)^T & (d_3^1)^T & 0^T & 0^T & 0^T & 0_{1\times 12} & 0_{1\times 3} \\ 0^T & (d_2^2)^T & 0^T & 0^T & 0^T & 0^T & (d_1^1)^T & 0^T & 0_{1\times 12} & 0_{1\times 3} \\ 0^T & 0^T & (d_3^2)^T & 0^T & 0^T & 0^T & 0^T & (d_2^1)^T & 0_{1\times 12} & 0_{1\times 3} \\ 0^T & 0^T & 0^T & (d_1^2)^T & 0^T & (d_3^1)^T & 0^T & 0^T & 0_{1\times 12} & 0_{1\times 3} \end{bmatrix} \tag{2.98}$$

The winch is connected to the trolley also by a revolute joint. The external constraint functions are written as

$$\boldsymbol{\Phi}_{\text{rev2}}(\overline{q}) = \begin{bmatrix} \boldsymbol{\varphi}^3 - \boldsymbol{\varphi}^2 \\ \boldsymbol{d}_2^2 \cdot \boldsymbol{d}_1^3 \\ \boldsymbol{d}_2^2 \cdot \boldsymbol{d}_3^3 \end{bmatrix} \tag{2.99}$$

and the corresponding constraint Jacobian is given by

$$\boldsymbol{G}_{\text{rev2}}(\overline{q}) = \begin{bmatrix} \boldsymbol{0}_{3\times12} & -\boldsymbol{I} & \boldsymbol{0}_{3\times3} & \boldsymbol{0}_{3\times3} & \boldsymbol{0}_{3\times3} & \boldsymbol{I} & \boldsymbol{0}_{3\times3} & \boldsymbol{0}_{3\times3} & \boldsymbol{0}_{3\times3} & \boldsymbol{0}_{3\times3} \\ \boldsymbol{0}_{1\times12} & \boldsymbol{0}^T & \boldsymbol{0}^T & (\boldsymbol{d}_1^3)^T & \boldsymbol{0}^T & \boldsymbol{0}^T & \boldsymbol{0}^T & \boldsymbol{0}^T & \boldsymbol{0}^T & \boldsymbol{0}_{1\times3} \\ \boldsymbol{0}_{1\times12} & \boldsymbol{0}^T & \boldsymbol{0}^T & (\boldsymbol{d}_3^3)^T & \boldsymbol{0}^T & \boldsymbol{0}^T & \boldsymbol{0}^T & \boldsymbol{0}^T & (\boldsymbol{d}_2^2)^T & \boldsymbol{0}_{1\times3} \end{bmatrix} \tag{2.100}$$

The payload is connected to the winch, and the constraint function is described by

$$\Phi = \left(\boldsymbol{\varphi}^4 - \boldsymbol{\varphi}^3\right) \cdot \left(\boldsymbol{\varphi}^4 - \boldsymbol{\varphi}^3\right) - (r_w\theta + l_0)^2 \tag{2.101}$$

in which the cable length is specified. The constraint Jacobian is then given by

$$\boldsymbol{G} = \left[\boldsymbol{0}_{1\times24} \;\; 2(\boldsymbol{\varphi}^3 - \boldsymbol{\varphi}^4)^T \;\; \boldsymbol{0}_{1\times9} \;\; 2(\boldsymbol{\varphi}^4 - \boldsymbol{\varphi}^3)^T \;\; 0 \;\; 0 \;\; -2r_w(r_w\theta + l_0)\right] \tag{2.102}$$

The additional coordinates $\overline{\theta}$, s and θ, shown in Fig. 2.9, are introduced into the rotationless formulation by applying a specific coordinate augmentation technique [87]. The variable $\overline{\theta}$ denotes the rotation angle of the girder bridge about the Z-axis relative to the X-axis, the variable s specifies the displacement of the trolley and the variable θ describes the rotation angle of the winch about the axis $\boldsymbol{d}_2^3$ of the body frame. In addition, in the inverse dynamics simulation the external force or torque, associated to the new coordinates $\overline{\theta}$, s and θ, can be incorporated into the rotatinless formulation as well.

The extended configuration vector is expressed by

$$\boldsymbol{q} = \begin{bmatrix} \overline{\boldsymbol{q}} \\ \overline{\theta} \\ s \\ \theta \end{bmatrix}_{42\times1} \tag{2.103}$$

The introduction of the three additional coordinates leads to three new constraints. The first constraint corresponding to $\overline{\theta}$ is expressed by

$$\Phi_{\text{aug1}}(\boldsymbol{q}) = \boldsymbol{d}_2^1 \cdot \boldsymbol{e}_1 + \sin\overline{\theta} + \boldsymbol{d}_2^1 \cdot \boldsymbol{e}_2 - \cos\overline{\theta} \tag{2.104}$$

and the constraint Jacobian is given by

$$G_{\text{aug1}}(q) = \begin{bmatrix} \mathbf{0}^T & \mathbf{0}^T & (e_1 + e_2)^T & \mathbf{0}^T & \mathbf{0}_{1\times 27} & (\cos\overline{\theta} + \sin\overline{\theta}) & 0 & 0 \end{bmatrix} \tag{2.105}$$

The second constraint corresponding to s is expressed by

$$\Phi_{\text{aug2}}(q) = d_1^1 \cdot \varphi^2 - s \tag{2.106}$$

which specifies the translational motion of the trolley. The constraint Jacobian is

$$G_{\text{aug2}}(q) = \begin{bmatrix} \mathbf{0}^T & \varphi^{2^T} & \mathbf{0}^T & \mathbf{0}^T & (d_1^1)^T & \mathbf{0}_{1\times 24} & 0 & -1 & 0 \end{bmatrix} \tag{2.107}$$

The third constraint corresponding to θ is expressed by

$$\Phi_{\text{aug3}}(q) = d_1^3 \cdot d_3^2 + \sin\theta + d_1^3 \cdot d_1^2 - \cos\theta \tag{2.108}$$

With regard to the treatment of the discretization, Equation (2.108) is rewritten in partitioned form

$$\Phi_{\text{aug3}}(q) = \Phi^1_{\text{aug3}}(\overline{q}) + \Phi^2_{\text{aug3}}(\theta) \tag{2.109}$$

with

$$\Phi^1_{\text{aug3}}(\overline{q}) = d_1^3 \cdot d_3^2 + d_1^3 \cdot d_1^2 \tag{2.110a}$$

$$\Phi^2_{\text{aug3}}(\theta) = \sin\theta - \cos\theta \tag{2.110b}$$

The corresponding constraint Jacobian is written as

$$G_{\text{aug3}}(q) = \begin{bmatrix} \mathbf{0}_{1\times 15} & (d_1^3)^T & \mathbf{0}^T & (d_1^3)^T & \mathbf{0}^T & (d_3^2 + d_1^2)^T & \mathbf{0}_{1\times 9} & 0 & 0 & (\cos\theta + \sin\theta) \end{bmatrix} \tag{2.111}$$

or in an alternative form

$$G_{\text{aug3}}(q) = \begin{bmatrix} G^1_{\text{aug3}}(\overline{q}) & G^2_{\text{aug3}}(\theta) \end{bmatrix} \tag{2.112}$$

with the components

$$G^1_{\text{aug3}}(\overline{q}) = \begin{bmatrix} \mathbf{0}_{1\times 15} & (d_1^3)^T & \mathbf{0}^T & (d_1^3)^T & \mathbf{0}^T & (d_3^2 + d_1^2)^T & \mathbf{0}_{1\times 9} & 0 & 0 \end{bmatrix} \tag{2.113a}$$

$$G^2_{\text{aug3}}(\theta) = \cos\theta + \sin\theta \tag{2.113b}$$

In summary, the formulation of the rotary crane model has 42 redundant coordinates, 18 internal constraints, 16 external constraints and 3 additional constraints. The degrees of freedom of the system is calculated by $f = 42 - 18 - 16 - 3 = 5$. It is to be noted that many external constraints are only linear and thus can be eliminated without destroying the structure of the DAEs (2.20). The reduction procedure can be found in Subsection 2.3.2.

Mass matrix and potential energy

After the application of the coordinate augmentation, the diagonal mass matrix of the system needs to be extended as:

$$\boldsymbol{M} = \begin{bmatrix} \boldsymbol{M}^1 & & & & & & \\ & \boldsymbol{M}^2 & & & & & \\ & & \boldsymbol{M}^3 & & & & \\ & & & \boldsymbol{M}^4 & & & \\ & & & & 0 & & \\ & & & & & 0 & \\ & & & & & & 0 \end{bmatrix}_{42\times 42} \tag{2.114}$$

where $\boldsymbol{M}^I (I = 1, 2, 3)$ is given in Equation (2.42) and

$$\boldsymbol{M}^4 = \begin{bmatrix} m_4 & & \\ & m_4 & \\ & & m_4 \end{bmatrix} \tag{2.115}$$

The potential energy of the rotary crane model is calculated by

$$V = m_1 \, g \, \boldsymbol{e}_3 \cdot \boldsymbol{\varphi}^1 + m_2 \, g \, \boldsymbol{e}_3 \cdot \boldsymbol{\varphi}^2 + m_3 \, g \, \boldsymbol{e}_3 \cdot \boldsymbol{\varphi}^3 + m_4 \, g \, \boldsymbol{e}_3 \cdot \boldsymbol{\varphi}^4 \tag{2.116}$$

The differential-algebraic equations (2.20) are obtained for the rotary crane model.

Reduced formulation of the rotary crane

The DAEs of the rotary crane model can be reduced by premultiplying the null space matrix. A mapping $\boldsymbol{F} : \mathbb{R}^f \mapsto \mathbb{R}^n$ can be found such that

$$q = \boldsymbol{F}(\boldsymbol{\mu}) \tag{2.117}$$

then the null space matrix $\boldsymbol{P}$ can be calculated through

$$\boldsymbol{P} = D\boldsymbol{F}(\boldsymbol{\mu}) \tag{2.118}$$

Following the reduction procedure in Subsection 2.3.2, the ODEs (2.17a)–(2.17b) of the rotary crane model can also be derived from the rotationless formulation.

Numerical discretization

For the direct discretization of the DAEs, the methodology developed by Gonzalez [45] is employed. This yields a specific second-order accurate algorithm called the basic energy-momentum (BEM) scheme [19], which is energy consistent and conserves momentum maps associated with symmetries of the underlying mechanical system.

A representative time interval $[t_n, t_{n+1}]$ with the time step $\Delta t = t_{n+1} - t_n$ is considered, and the state space coordinates $\boldsymbol{q}_n \in Q$, $\boldsymbol{v}_n \in \mathbb{R}^n$ at t_n are given.

Then the discretized version of (2.20a)–(2.20c) is given by

$$\boldsymbol{q}_{n+1} - \boldsymbol{q}_n = \frac{\Delta t}{2}(\boldsymbol{v}_n + \boldsymbol{v}_{n+1}) \tag{2.119a}$$

$$\boldsymbol{M}(\boldsymbol{v}_{n+1} - \boldsymbol{v}_n) = -\Delta t \left[\overline{\nabla} V(\boldsymbol{q}_n, \boldsymbol{q}_{n+1}) + \boldsymbol{G}^T(\boldsymbol{q}_n, \boldsymbol{q}_{n+1})\boldsymbol{\lambda}_{n,n+1}\right] \tag{2.119b}$$

$$\boldsymbol{\Phi}(\boldsymbol{q}_{n+1}) = \boldsymbol{0} \tag{2.119c}$$

where the discrete Lagrange multipliers $\boldsymbol{\lambda}_{n,n+1}$ are assumed to be constant in the time interval $[t_n, t_{n+1}]$.

The advantageous algorithmic conservation properties of the BEM scheme are linked to the discrete gradient of a function $\mathcal{F}$: $\mathbb{R}^n \mapsto \mathbb{R}$. If $\mathcal{F}$ is at most quadratic, then the discrete gradient is identical to the standard gradient, which is evaluated in the mid-point configuration $\boldsymbol{q}_{n+\frac{1}{2}}$. This implies that $\overline{\nabla}\mathcal{F}(\boldsymbol{q}_n, \boldsymbol{q}_{n+1}) = \nabla\mathcal{F}(\boldsymbol{q}_{n+\frac{1}{2}})$. In Equation (2.119b) the discrete gradient is applied to the potential energy function V, i.e. $\overline{\nabla} V(q_n, q_{n+1})$, and to the constraint functions $\boldsymbol{\Phi}$. The discrete constraint Jacobian of $\boldsymbol{\Phi}$ is expressed by

$$G^T(q_n, q_{n+1}) = \left[\overline{\nabla}\Phi_1(q_n, q_{n+1}), \overline{\nabla}\Phi_2(q_n, q_{n+1}), \cdots, \overline{\nabla}\Phi_m(q_n, q_{n+1})\right] \tag{2.120}$$

Considering the constraints of the rotary crane example again, the associated discrete gradient coincides with the mid-point evaluation of the continuous constraint Jacobian, since most of the constraints are at most quadratic. In contrast, the additional constraint functions require a special treatment. Taking the third additional constraint function $\Phi_{\text{aug3}}(q)$ as an example, the discrete gradient is expressed as

$$G_{\text{aug3}}(q_n, q_{n+1}) = \left[G^1_{\text{aug3}}(\overline{q}_{n+\frac{1}{2}}) \quad G^2_{\text{aug3}}(\theta_n, \theta_{n+1})\right] \tag{2.121}$$

with the component

$$G^2_{\text{aug3}}(\theta_n, \theta_{n+1}) = \frac{\Phi^2_{\text{aug3}}(\theta_{n+1}) - \Phi^2_{\text{aug3}}(\theta_n)}{\theta_{n+1} - \theta_n} \tag{2.122}$$

If $\theta_{n+1} \to \theta_n$, then $G^2_{\text{aug3}}(\theta_n, \theta_{n+1}) \to (\Phi^2_{\text{aug3}}(\theta_n))'$, where the constraint derivative is given by $(\Phi^2_{\text{aug3}}(\theta))' = \cos\theta + \sin\theta$.

To solve the discretized equations of motion (2.119a)–(2.119c), inserting v_{n+1} calculated from Equation (2.119a) into (2.119b) leads to a system of nonlinear algebraic equations, which can be solved for the $n + m$ unknowns $(q_{n+1}, \lambda_{n,n+1})$. Thus, $q_{n+1} \in Q$, $v_{n+1} \in \mathbb{R}^n$, $\lambda_{n,n+1} \subset \mathbb{R}^m$ can be determined by applying the BEM scheme, which is discussed thoroughly in [9].

Forward dynamics simulation

Next the simulation[4] of the rotary crane now using the redundant coordinates is repeated. The simulation will focus on the algorithmic conservation properties of the BEM scheme.

[4] The origin of the inertial Cartesian coordinate system XYZ is placed at the half height position of the pillar rather than at the top of the pillar (see Fig. 2.9). The placement of the origin of the inertial Cartesian coordinate system has no influence on the rotationless formulation.

In the simulation, the mass, value of Euler tensor and dimension of each rigid body of the rotary crane are listed in Table 2.2. The principal values of the Euler tensor E_i can be calculated by

$$E_i = \frac{1}{2}(J_j + J_k - J_i) \tag{2.123}$$

for even permutations of the indices (i, j, k) and the principal values of the classical (convected) inertia tensor are given by J_i, J_j, J_k. Moreover, the initial configuration is defined by: $\overline{\theta} = 0$, $s = 2.5\,\text{m}$, $\theta = 0$, $l_0 = 5\,\text{m}$. The initial angular velocity of the first body is given by $\dot{\overline{\theta}} = 1\,\text{m/s}$. During the simulation there are no other external forces and torques except the gravitational force.

body	m [kg]	E_1 [kg · m^2]	E_2	E_3	length [m]	width	depth
1	100	8.33	8.33	208.33	5	1	1
2	50	1.04	1.04	1.04	0.5	0.5	0.5
3	3	0.01	0.25	0.01	1	0.2	0.2
4	10	—	—	—	—	—	—

Table 2.2: Data of mass, Euler tensor and dimension for each body of the rotary crane.

The rotary crane system can be classified as an autonomous Hamiltonian system with symmetry. As shown in Fig. 2.10, the total energy and the third component of the angular momentum L_z are conserved quantities. Some snapshots of the simulated motion of the rotary crane model are illustrated in Fig. 2.11. Obviously, the numerical results and the snapshots are identical to those obtained from the generalized coordinates formulation. It is to be noted that the present energy-momentum scheme conserves the quantities independent of the time step size.

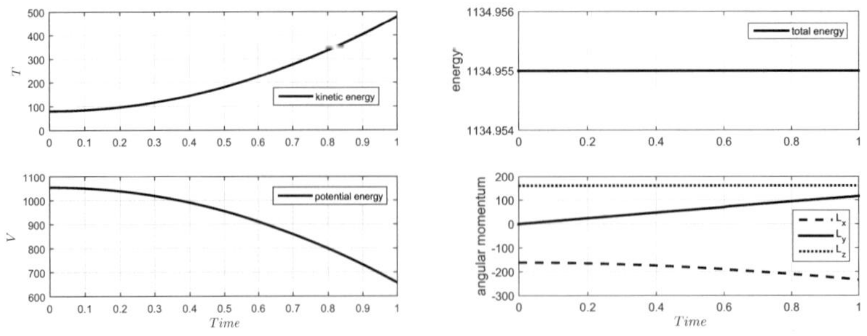

Figure 2.10: Rotationless formulation: Energy and angular momentum of the rotary crane with the time step of $\Delta t = 10^{-2}\,\text{s}$.

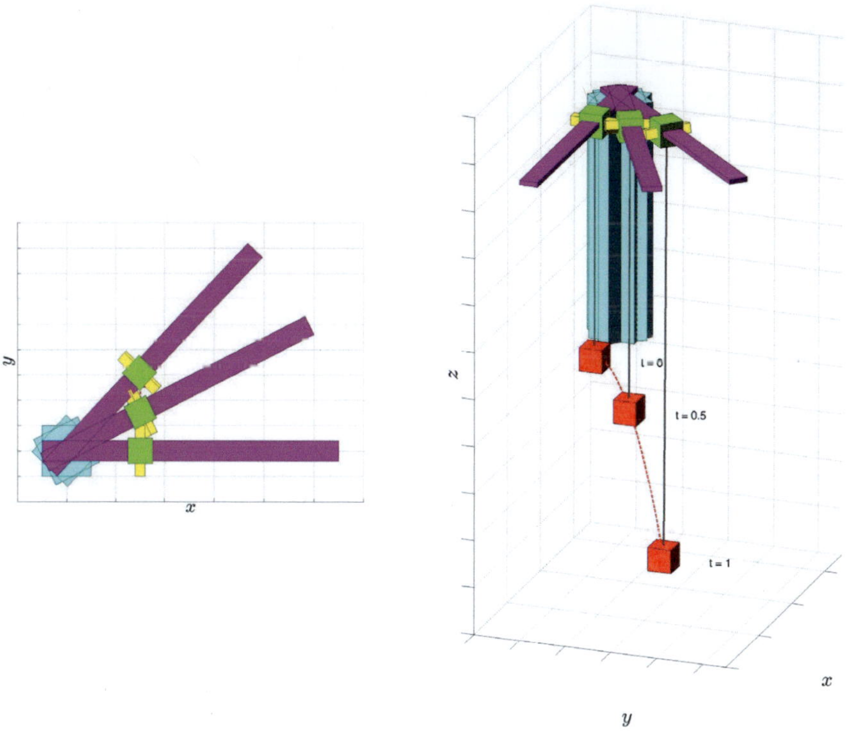

Figure 2.11: Rotationless formulation: Snapshots of the simulation of the rotary crane at specific time points.

3 Numerical integration schemes

In Chapter 2 the equations of motion of mechanical systems have been derived by applying the Lagrange's equations. The equations of motion pertaining to constrained mechanical systems are index-3 differential-algebraic equations, which can be recapitulated here:

$$\dot{\boldsymbol{q}} - \boldsymbol{v} = \mathbf{0} \tag{3.1a}$$

$$\boldsymbol{M}\dot{\boldsymbol{v}} + \nabla V(\boldsymbol{q}) + \boldsymbol{G}^T(\boldsymbol{q})\boldsymbol{\lambda} = \mathbf{0} \tag{3.1b}$$

$$\boldsymbol{\Phi}(\boldsymbol{q}) = \mathbf{0} \tag{3.1c}$$

Then the numerical time-stepping schemes are applied to solve the above continuous equations of motion (3.1a)–(3.1c). For the time discretization, a representative time interval $\mathbb{I}_n = [t_n, t_{n+1}]$ with the time step $\Delta t = t_{n+1} - t_n$ is considered, and the coordinates $\boldsymbol{q}_n \in Q$, velocities $\boldsymbol{v}_n \in \mathbb{R}^n$ and the Lagrange multipiers $\boldsymbol{\lambda}_n \in \mathbb{R}^m$ at the time node t_n are given. The task is to compute the coordinates $\boldsymbol{q}_{n+1} \in Q$, velocities $\boldsymbol{v}_{n+1} \in \mathbb{R}^n$ and the Lagrange multipliers $\boldsymbol{\lambda}_{n+1} \in \mathbb{R}^m$ at the next time node t_{n+1}.

Some numerical integration schemes, which are usually used for the direct time discretization of the underlying DAEs, will be outlined in this chapter.

3.1 Implicit Euler method

The implicit Euler method is a basic time-stepping scheme for the numerical integration of DAEs. The application of the implict Euler method leads to the direct time discretization of the index-3 DAEs (3.1a)–(3.1c):

$$
\begin{aligned}
q_{n+1} - q_n &= \Delta t \, v_{n+1} &\quad (3.2a)\\
M\,(v_{n+1} - v_n) &= -\Delta t \left[\nabla V(q_{n+1}) + G^T(q_{n+1}) \lambda_{n+1} \right] &\quad (3.2b)\\
\Phi\,(q_{n+1}) &= 0 &\quad (3.2c)
\end{aligned}
$$

Then the Newton-Raphson method can be applied to solve the resulting nonlinear algebraic equations (3.2a)–(3.2c). The details about the Newton-Raphson method can be found in [40]. Accordingly, the coordinates, velocities $q_{n+1} \in Q$, $v_{n+1} \in \mathbb{R}^n$ as well as the Lagrange multipliers $\lambda_{n+1} \in \mathbb{R}^m$ at the time node t_{n+1} can be obtained.

It is worth noting that the implicit Euler method is first order accurate and exhibits numerical damping. Due to the numerical damping the total energy of a conservative mechanical system is not conserved, but decays with the time during the numerical simulation.

3.2 Mid-point-type rule

The mid-point-type rule has been applied to discretize the ordinary differential equations in the example of the rotary crane. It can also be used for the time discretization of the differential-algebraic equations (3.1a)–(3.1c). Then the set of discretized equations is expressed as

$$
\begin{aligned}
q_{n+1} - q_n &= \Delta t \, v_{n+\frac{1}{2}} &\quad (3.3a)\\
M\,(v_{n+1} - v_n) &= -\Delta t \left[\nabla V(q_{n+\frac{1}{2}}) + G^T(q_{n+\frac{1}{2}}) \lambda_{n,n+1} \right] &\quad (3.3b)\\
\Phi\,(q_{n+1}) &= 0 &\quad (3.3c)
\end{aligned}
$$

in which the Lagrange multipliers $\lambda \in \mathbb{R}^m$ are approximated by constant values $\lambda_{n,n+1} \in \mathbb{R}^m$ during the time interval $\mathbb{I}_n$. This approximation leads to possible discontinuities on the boundaries of the time interval $\mathbb{I}_n$. Besides, the geometric constraint conditions need to be satisfied at each time node t_n and t_{n+1}. It is to be noted that in nonlinear elastodynamics the mid-point rule does not conserve the total energy of the mechanical system.

3.3 Energy-momentum scheme

The energy-momentum scheme is a structure-preserving time-stepping scheme, which was originally proposed by Simo and Tarnow [85] in the context of nonlinear elastodynamics. It inherits conservation of energy and momentum maps and has also been applied to rigid body dynamics, in which the rotationless formulation of multibody systems benefits the design of the energy-momentum scheme [19]. The application of the energy-momentum scheme in various problems can be found for example in [12, 14, 18, 23, 24, 45, 49].

The energy-momentum scheme has already been applied to discretize the differential-algebraic equations in the example of the rotary crane. The set of discretized equations is recapitulated:

$$q_{n+1} - q_n = \frac{\Delta t}{2}(v_n + v_{n+1}) \tag{3.4a}$$

$$\boldsymbol{M}\,(v_{n+1} - v_n) = -\Delta t\left[\overline{\nabla} V(q_n, q_{n+1}) + \boldsymbol{G}^T(q_n, q_{n+1})\boldsymbol{\lambda}_{n,n+1}\right] \tag{3.4b}$$

$$\boldsymbol{\Phi}\,(q_{n+1}) = \boldsymbol{0} \tag{3.4c}$$

where the discrete constraint Jacobian is defined by

$$\boldsymbol{G}^T(q_n, q_{n+1}) = \left[\overline{\nabla}\Phi_1(q_n, q_{n+1}), \overline{\nabla}\Phi_2(q_n, q_{n+1}), \cdots, \overline{\nabla}\Phi_m(q_n, q_{n+1})\right] \tag{3.5}$$

The discrete gradient (derivative) denoted by $\overline{\nabla}$ is crucial to the algorithmic conservation of both energy and angular momentum. It satisfies the directionality property

$$\overline{\nabla} f(q_n, q_{n+1}) \cdot (q_{n+1} - q_n) = f(q_{n+1}) - f(q_n) \tag{3.6}$$

which is of key importance for algorithmic energy conservation. It is worth mentioning that if the function $\mathcal{F} : \mathbb{R}^n \mapsto \mathbb{R}$ is at most quadratic then the discrete gradient coincides with the mid-point evaluation of the standard gradient. That is

$$\overline{\nabla}\mathcal{F}(q_n, q_{n+1}) = \nabla\mathcal{F}(q_{n+\frac{1}{2}}) \tag{3.7}$$

An in-depth investigation of properties of discrete derivatives can be found in [44].

In contrast to the implicit Euler method, the energy-momentum scheme is second order accurate and exhibits superior numerical stability. The drawback of both schemes lies in the condition number of the interation matrix for the solution of the nonlinear system of equations (3.4a)–(3.4c). It implies that the condition number of the iteration matrix becomes more and more ill-conditioned for decreasing time steps. To remedy this drawback, a further modification can be accomplished by applying the discrete null space method and the reparametrization of the remaining unknows. Application of the two size-reduction steps yields the so-called reduced energy-momentum scheme. Further details about the reduction method can be found in [7, 8, 9, 13, 79, 87].

3.4 Variational integrator

The concept of a variational integrator is based on a direct discretization of the action integral, whereas the previous time-stepping schemes rely on the direct discretization of the underlying equations of motion. In this section specific variational integrators will be derived according to the lecture notes from Betsch [11]. For this purpose, it is assumed that the time interval $\mathbb{I} = [0, T]$ is divided into N equidistant intervals $\mathbb{I}_n = [t_n, t_{n+1}]$ with the time step $\Delta t = t_{n+1} - t_n$.

Minimal coordinates

Here the configuration vector q is used to represent the minimal coordinates. The action integral is then defined as

$$S = \int_0^T L(q, \dot{q})\,\mathrm{d}t = \sum_{n=0}^{N-1} \int_{t_n}^{t_{n+1}} L(q, \dot{q})\,\mathrm{d}t \tag{3.8}$$

The discrete Lagrangian is introduced to approximate the integral $\int_{t_n}^{t_{n+1}} L(q, \dot{q})\,\mathrm{d}t$. In this case there is

$$L_d(q_n, q_{n+1}) \approx \int_{t_n}^{t_{n+1}} L(q, \dot{q})\,\mathrm{d}t \tag{3.9}$$

Accordingly, the discrete action integral is expressed by

$$S_d(q_0, \dots, q_N) = \sum_{n=0}^{N-1} L_d(q_n, q_{n+1}) \tag{3.10}$$

The application of the discrete variational principle requires that $\delta S_d = 0$ with the boundary points q_0 and q_N held fixed. This yields

$$\delta S_d = \delta \sum_{n=0}^{N-1} L_d(q_n, q_{n+1}) = \sum_{n=0}^{N-1} \delta L_d(q_n, q_{n+1}) = 0 \tag{3.11}$$

By using the Gâteaux derivative (see Appendix A.1), the partial derivatives[1] can be calculated and the variation of the discrete action integral reads

$$\begin{aligned}
\delta S_d &= \sum_{n=0}^{N-1} \delta L_d(q_n, q_{n+1}) = \sum_{n=0}^{N-1} \left[D_1 L_d(q_n, q_{n+1}) \cdot \delta q_n + D_2 L_d(q_n, q_{n+1}) \cdot \delta q_{n+1} \right] \\
&= D_1 L_d(q_0, q_1) \cdot \delta q_0 + \sum_{n=1}^{N-1} \delta q_n \cdot \left[D_1 L_d(q_n, q_{n+1}) + D_2 L_d(q_{n-1}, q_n) \right] + \\
&\quad D_2 L_d(q_{N-1}, q_N) \cdot \delta q_N \\
&= \sum_{n=1}^{N-1} \delta q_n \cdot \left[D_1 L_d(q_n, q_{n+1}) + D_2 L_d(q_{n-1}, q_n) \right] = 0
\end{aligned} \tag{3.12}$$

where a discrete integration by parts [71] is used and the end point conditions $\delta q(t_0) = \delta q_0 = 0$ and $\delta q(t_N) = \delta q_N = 0$ are considered. It is required that the variation of the discrete action integral is equal to zero for any choice of δq_n. This leads to the discrete Euler-Lagrange (DEL) equations:

$$D_2 L_d(q_{n-1}, q_n) + D_1 L_d(q_n, q_{n+1}) = 0 \tag{3.13}$$

which must hold for $n = 1, \ldots, N-1$.

The trapezoidal rule is used to approximate the integral in Equation (3.9), which can be expressed by

$$L_d(q_n, q_{n+1}) = \frac{\Delta t}{2} \left[L\left(q_n, \frac{q_{n+1} - q_n}{\Delta t}\right) + L\left(q_{n+1}, \frac{q_{n+1} - q_n}{\Delta t}\right) \right] \tag{3.14}$$

1 Here the notations $D_1 L_d(x, y) = \dfrac{\partial L_d}{\partial x}$ and $D_2 L_d(x, y) = \dfrac{\partial L_d}{\partial y}$ are used.

by using the following expressions

$$q\,|_{\mathbb{I}_n} = q_n + \frac{t - t_n}{\Delta t}(q_{n+1} - q_n)$$
$$\dot{q}\,|_{\mathbb{I}_n} = \frac{(q_{n+1} - q_n)}{\Delta t} \tag{3.15}$$

The continuous Lagrangian of the mechanical system under consideration can be given by

$$L(q, \dot{q}) = \frac{1}{2}\dot{q} \cdot \boldsymbol{M}\dot{q} - V(q) \tag{3.16}$$

With regard to Equation (3.14), the discrete Lagrangian is thus written as

$$L_d(q_n, q_{n+1}) = \frac{1}{2\Delta t}\left(q_{n+1} - q_n\right) \cdot \boldsymbol{M}\left(q_{n+1} - q_n\right) - \frac{\Delta t}{2}\left[V\left(q_n\right) + V\left(q_{n+1}\right)\right] \tag{3.17}$$

Then the discrete Euler-Lagrange equations in (3.13) are applied and the partial derivatives are

$$D_1 L_d(q_n, q_{n+1}) = -\boldsymbol{M}\frac{q_{n+1} - q_n}{\Delta t} - \frac{\Delta t}{2}\nabla V\left(q_n\right) \tag{3.18a}$$
$$D_2 L_d(q_n, q_{n+1}) = \boldsymbol{M}\frac{q_{n+1} - q_n}{\Delta t} - \frac{\Delta t}{2}\nabla V\left(q_{n+1}\right) \tag{3.18b}$$

Equation (3.18b) can also be expressed by

$$D_2 L_d(q_{n-1}, q_n) = \boldsymbol{M}\frac{q_n - q_{n-1}}{\Delta t} - \frac{\Delta t}{2}\nabla V\left(q_n\right) \tag{3.19}$$

Inserting Equation (3.19) and (3.18a) into Equation (3.13) leads to the following non-linear algebraic equations

$$\frac{1}{\Delta t}\boldsymbol{M}\left(-q_{n+1} + 2q_n - q_{n-1}\right) - \Delta t \nabla V\left(q_n\right) = 0 \tag{3.20}$$

and an alternative form is given by

$$\frac{1}{(\Delta t)^2}\boldsymbol{M}\left(q_{n+1} - 2q_n + q_{n-1}\right) = -\nabla V\left(q_n\right) \tag{3.21}$$

in which the acceleration has been approximated through the second order central difference, that is

$$\ddot{q} = \frac{q_{n+1} - 2q_n + q_{n-1}}{(\Delta t)^2} \tag{3.22}$$

If the coordinates q_{n-1} at the time node t_{n-1} and q_n at t_n are already known, then the coordinates q_{n+1} at t_{n+1} can be calculated. It is obvious that the variational integrator is an explicit two-step method, that is, $(q_{n-1}, q_n) \rightarrow (q_n, q_{n+1})$. Nevertheless, it can be implemented as an one-step method by introducing the discrete momentum

$$P_n = -D_1 L_d(q_n, q_{n+1}) \tag{3.23}$$

Due to Equation (3.13), the discrete momentum is also given by

$$P_n = D_2 L_d(q_{n-1}, q_n) \tag{3.24}$$

or in an alternative form

$$P_{n+1} = D_2 L_d(q_n, q_{n+1}) \tag{3.25}$$

In this case, if q_n and P_n are given, then q_{n+1} can be calculated from Equation (3.23). Then the momentum P_{n+1} can be obtained from Equation (3.25). The procedure is expressed by $(q_n, P_n) \rightarrow (q_{n+1}, P_{n+1})$. Instead of the trapezoidal rule, the mid-point rule can also be used to approximate the integral in Equation (3.9)

$$L_d(q_n, q_{n+1}) = \Delta t L\left(\frac{q_{n+1} + q_n}{2}, \frac{q_{n+1} - q_n}{\Delta t}\right) \tag{3.26}$$

Other descriptions of the variational integrator can be found in [22, 68, 66, 71].

Redundant coordinates

Here the configuration vector q is used to represent the redundant coordinates. According to the variational symplectic-momentum integrator proposed by Leyendecker et al. [68] (see also [22]), for constrained mechanical systems, the discretization of the extended action integral (2.11) with $t_1 = 0$ and $t_2 = T$ yields the discrete extended action integral

$$\widetilde{S}_\mathrm{d} = \sum_{n=0}^{N-1} \left[L_\mathrm{d}(q_n, q_{n+1}) - \frac{\Delta t}{2} \left(\lambda_n \cdot \Phi(q_n) + \lambda_{n+1} \cdot \Phi(q_{n+1}) \right) \right] \tag{3.27}$$

with the expression

$$\int_0^T \lambda \cdot \Phi(q) \, \mathrm{d}t \approx \frac{\Delta t}{2} \left(\lambda_n \cdot \Phi(q_n) + \lambda_{n+1} \cdot \Phi(q_{n+1}) \right) \tag{3.28}$$

Similarly, the variational principle requires that $\delta \widetilde{S}_d = 0$ for fixed end points q_0 and q_N. This yields

$$\delta q_0 \cdot \left[D_1 L_d(q_0, q_1) - \frac{1}{2} D\Phi_d^T(q_0)\,\lambda_0 \right] + \delta q_N \cdot \left[D_2 L_d(q_{N-1}, q_N) - \frac{1}{2} D\Phi_d^T(q_N)\,\lambda_N \right]$$
$$+ \sum_{n=1}^{N-1} \delta q_n \cdot \left[D_1 L_d(q_n, q_{n+1}) + D_2 L_d(q_{n-1}, q_n) - D\Phi_d^T(q_n)\,\lambda_n \right] = 0 \tag{3.29}$$

for any choice of δq_n with $0 \leqslant n \leqslant N$ and the end point conditions $\delta q_0 = \delta q_N = 0$, and the condition

$$\frac{1}{2}\delta\lambda_0 \cdot \Phi_d(q_0) + \frac{1}{2}\delta\lambda_N \cdot \Phi_d(q_N) + \sum_{n=1}^{N-1} \delta\lambda_n \cdot \Phi_d(q_n) = 0 \tag{3.30}$$

for any choice of λ_n with $0 \leqslant n \leqslant N$. Here the abbreviation $\Phi_d(q) = \Delta t\, \Phi(q)$ is used. Eventually, the discrete Euler-Lagrange (DEL) equations for constrained mechanical systems read

$$D_2 L_d(q_{n-1}, q_n) + D_1 L_d(q_n, q_{n+1}) - G_d^T(q_n)\,\lambda_n = 0 \tag{3.31a}$$
$$\Phi(q_{n+1}) = 0 \tag{3.31b}$$

where $G_d(q) = \Delta t\, D\Phi(q)$. In view of the continuous Lagrangian in Equation (3.16), the discrete Lagrangian in Equation (3.14) is given by

$$L_d(q_n, q_{n+1}) = \frac{1}{2\Delta t}\,(q_{n+1} - q_n) \cdot M\,(q_{n+1} - q_n) - \Delta t V\left(\frac{q_{n+1} + q_n}{2}\right) \tag{3.32}$$

Applying the discrete Euler-Lagrange equations (3.31a)–(3.31b) leads to

$$M\frac{(q_{n+1} - 2q_n + q_{n-1})}{\Delta l} + \frac{\Delta t}{2}\left[\nabla V(q_{n-\frac{1}{2}}) + \nabla V(q_{n+\frac{1}{2}})\right] + G_d^T(q_n)\,\lambda_n = 0 \tag{3.33a}$$
$$\Phi(q_{n+1}) = 0 \tag{3.33b}$$

with the expressions

$$q_{n-\frac{1}{2}} = \frac{1}{2}(q_{n-1} + q_n), \quad q_{n+\frac{1}{2}} = \frac{1}{2}(q_n + q_{n+1}) \tag{3.34}$$

The discrete momentum can be defined by

$$
\begin{aligned}
\boldsymbol{P}_n^- &= -D_1 L_\mathrm{d}(\boldsymbol{q}_n, \boldsymbol{q}_{n+1}) + \frac{1}{2}\mathbf{G}_\mathrm{d}^T(\boldsymbol{q}_n) \\
\boldsymbol{P}_n^+ &= -D_2 L_\mathrm{d}(\boldsymbol{q}_{n-1}, \boldsymbol{q}_n) - \frac{1}{2}\mathbf{G}_\mathrm{d}^T(\boldsymbol{q}_n)
\end{aligned}
\tag{3.35}
$$

Accordingly, the DEL equations can be written in an alternative form

$$
\boldsymbol{P}_n^+ - \boldsymbol{P}_n^- = 0 \tag{3.36}
$$

3.5 Andrew's squeezer mechanism

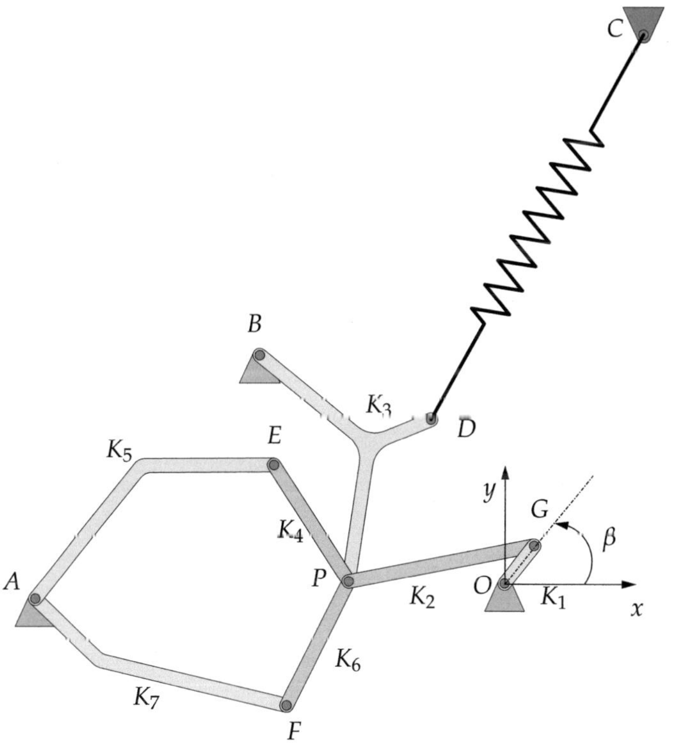

Figure 3.1: Andrew's squeezer mechanism: Setup.

A classical benchmark problem for multibody dynamics is used to demonstrate the application of numerical integration schemes. In this benchmark problem the rotationless formulation (see Section 2.3) is used. According to [91, Sec. 3.6.9] this mechanism can be traced back to the PhD thesis by G.C. Andrews (1971). A detailed description of Andrews' squeezer mechanism can be found in the Multibody Systems Handbook [80], the book [50, Ch. VII.7] and the Technical Report [55]. Related numerical investigations have also been documented in [4, Sec. 5.2].

As shown in Fig. 3.1, the multibody system at hand consists of 7 rigid bodies interconnected by frictionless revolute joints. The coordinates of the joints are given in Table 3.1. In addition, the inertia properties as well as the coordinates of the center of mass for all bodies are given in Table 3.2.

Moreover a spring with spring coefficient $c_0 = 4530\,\mathrm{N/m}$ and unstretched length $l_0 = 7.785 \times 10^{-2}\,\mathrm{m}$ is connected to the present multibody system. The spring length in the initial configuration ($t_0 = 0$) is $5.267 \times 10^{-2}\,\mathrm{m}$.

The body-fixed frames are located in the center of mass of each body. The mechanism is driven by a motor located at point O. A constant torque $M = 0.033\,\mathrm{N} \cdot \mathrm{m}$ is applied. In the initial configuration ($t_0 = 0$) the mechanism is at rest. Obviously, the mechanism at hand has one degree of freedom. In the numerical simulations gravitation is not taken into account.

In Fig. 3.2 the two displacement components corresponding to joint (P) are plotted versus time. Similarly the angle β is plotted over time in Fig. 3.3. Again it can be observed from Fig. 3.4 that the EM scheme adheres to the balance law for energy, whereas both Gen-α[2] and VI fail to satisfy this balance law. The situation is shown in more detail in Fig. 3.5. Of course, since Gen-α and VI are consistent, refinement of the time step yields an improved fulfillment of the balance of energy. This is shown in Figs. 3.6 and 3.7. Finally, to illustrate the motion of the whole multibody system at hand several snapshots are plotted in Fig. 3.8.

[2] More details about the generalized-α scheme can be found for example in [4].

Joints	x [m]	y [m]
O	0	0
A	-0.06934	-0.00227
B	-0.03635	0.03273
C	0.01400	0.07200
D	-0.01047	0.02536
E	-0.03400	0.01646
F	-0.03163	-0.01562
G	0.00699	-0.00043
P	-0.02096	0.00130

Table 3.1: Andrew's squeezer mechanism. Coordinates of the joints.

Link	Mass [kg]	Rotational inertia [$kg \cdot m^2$]	x [m]	y [m]
1	0.04325	$2.194 \cdot 10^{-6}$	$9.182 \cdot 10^{-4}$	$5.700 \cdot 10^{-5}$
2	0.00365	$4.410 \cdot 10^{-7}$	$-4.491 \cdot 10^{-3}$	$2.788 \cdot 10^{-4}$
3	0.02373	$5.255 \cdot 10^{-6}$	$1.874 \cdot 10^{-2}$	$2.048 \cdot 10^{-2}$
4	0.00706	$5.667 \cdot 10^{-7}$	$-3.022 \cdot 10^{-2}$	$1.207 \cdot 10^{-2}$
5	0.07050	$1.169 \cdot 10^{-5}$	$-5.324 \cdot 10^{-2}$	$1.663 \cdot 10^{-2}$
6	0.00706	$5.667 \cdot 10^{-7}$	$-2.854 \cdot 10^{-2}$	$-1.072 \cdot 10^{-2}$
7	0.05498	$1.912 \cdot 10^{-5}$	$-5.926 \cdot 10^{-2}$	$-1.060 \cdot 10^{-2}$

Table 3.2: Andrew's squeezer mechanism: Inertia data and coordintates of the center of mass.

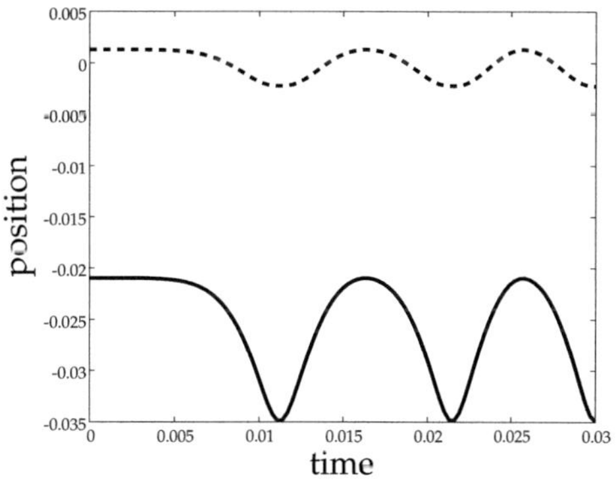

Figure 3.2: Displacement of hinge (P).

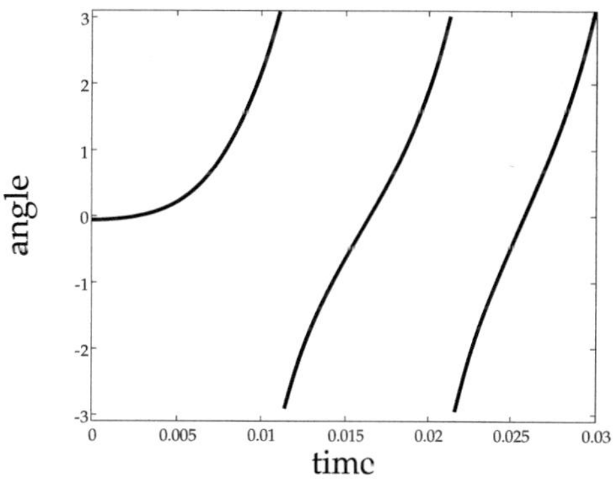

Figure 3.3: Angle β [rad].

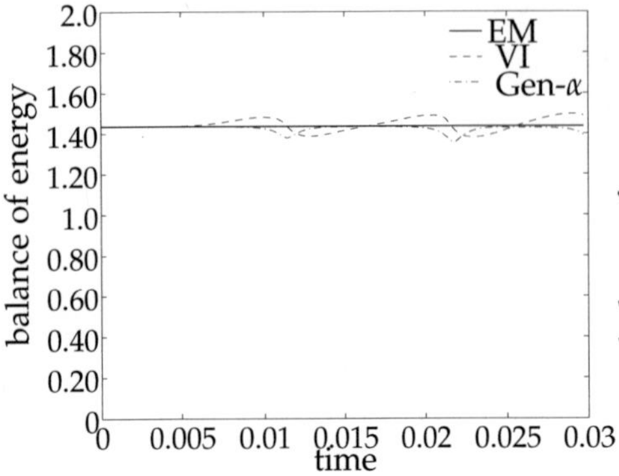

Figure 3.4: Time step $\Delta t = 3 \cdot 10^{-4}$s.

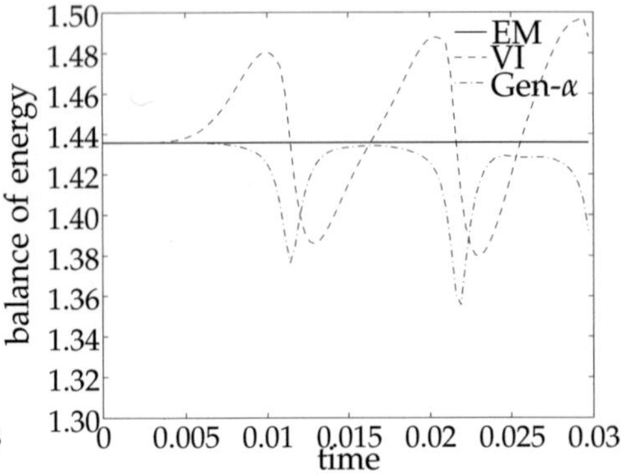

Figure 3.5: Time step $\Delta t = 3 \cdot 10^{-4}$s.

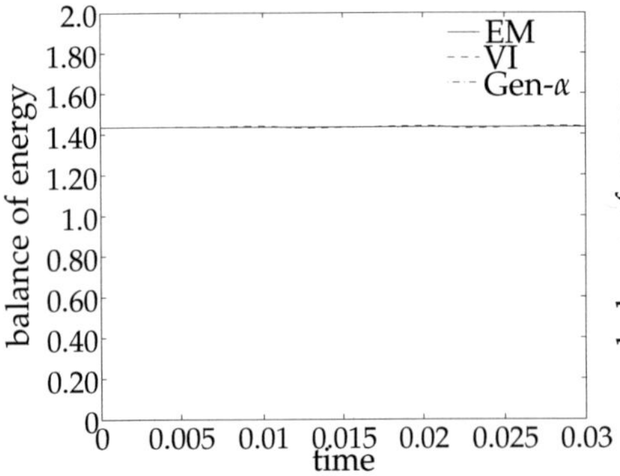

Figure 3.6: Time step $\Delta t = 3 \cdot 10^{-5}$s.

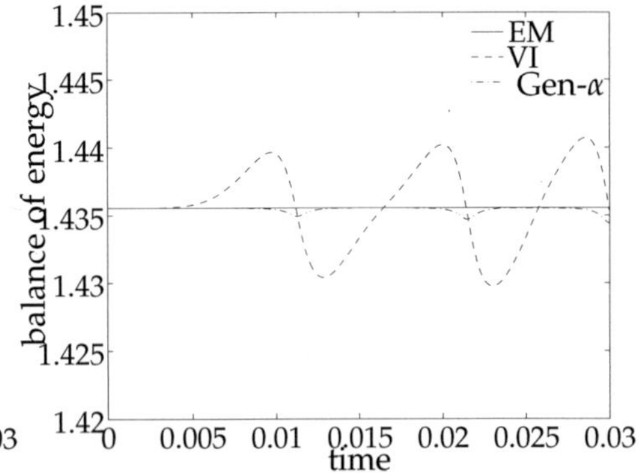

Figure 3.7: Time step $\Delta t = 3 \cdot 10^{-5}$s.

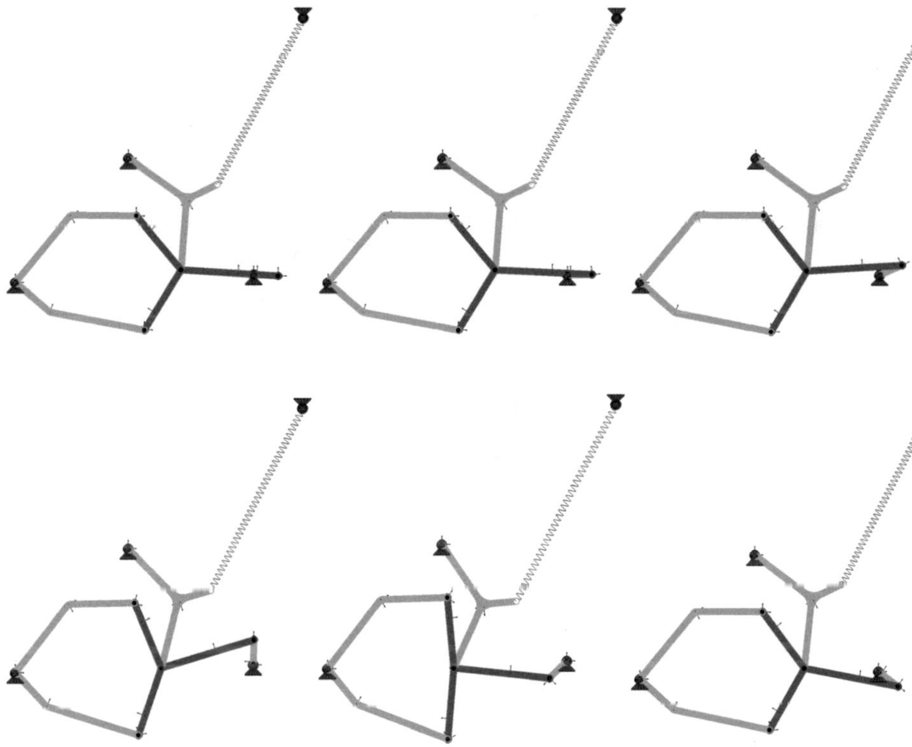

Figure 3.8: Andrew's squeezer mechanism: Snapshots at $t = \{0, 3, 6, 9, 12, 15\}$ ms.

4 Inverse dynamics simulation of multibody systems

The inverse dynamics control problem can be stated as follows: given a desired or prescribed motion of a mechanical system, determine the control inputs that force the system to complete this specified motion, and the determination is based on the dynamic model of the controlled mechanical system [29]. The derivation of the dynamic model of the controlled mechanical system can be achieved by either generalized coordinates formulation in terms of minimal coordinates or rotationless formulation in terms of redundant coordinates (see Chapter 2). The method of inverse dynamics is often most suitable for the trajectory tracking control problems of multibody systems. Depending on the ratio of the number a of independent control inputs to the number f of degrees of freedom, the multibody systems fall into three main categories: overactuated $(a > f)$, fully actuated $(a = f)$ and underactuated $(a < f)$ mechanical systems [26].

Overactuated systems have more control inputs than degrees of freedom and are ofen found in aerospace, automotive, robotics applications and biomechanical models. Fully actuated systems have as many control inputs/outputs as degrees of freedom and the motion of the systems is fully specified by the task requirements. Given a fully prescribed motion of a system at the position, velocity and acceleration levels, the desired feedforward control law is the consequent result of a pure algebraic resolution of the dynamic equations [26]. The inverse dynamics control of this type has been intensively described in robotic textbooks, such as Craig [36], Murray et al. [75], Spong et al. [86] (see Seifried [81]). Moreover, the previously developed structure-preserving integrators for index-3 DAEs have been applied successfully (see Uhlar and Betsch [89]), where servo constraints are incorporated into an energy-momentum scheme emanating from the direct discretization of the underlying DAEs.

In fully actuated and overactuated multibody systems, all degrees of freedom can explicitly be regulated by available controls, such as control forces and torques. This is not the case for underactuated multibody systems, which have fewer control inputs than degrees of freedom. The underactuation makes the inverse dynamics control problem more challenging. Moreover, for trajectory tracking control of underactuated systems required to complete a partly specified motion, an accurate and efficient feedforward control law is often necessary. Therefore, the main focus is on the inverse dynamics simulation of underactuated mechanical systems, which relies on servo constraints for the partial specification of the motion of the systems.

4.1 Underactuated mechanical systems with servo constraints

In underactuated mechanical systems, the number a of control inputs, equal to the number of control outputs, which are in general desired performance goals of a dynamical system, is smaller than the number f of degrees of freedom, $a < f$ [29]. Typical examples of underactuated mechanical systems are cranes like overhead cranes, rotary cranes, and flexible multibody systems like manipulators with flexible joints or members [54, 69]. Some other examples of underactuated systems can be found in [25, 27].

A possible performance task of underactuated systems is the output trajectory tracking, for example, the trajectory tracking of the end-effector of manipulators. Thus, the main focus here is on the specification of trajectories of specific points of a multibody system such as the end-effector of a robot. In this connection, the desired system outputs can be described in terms of the system states and modeled as servo constraints [56, 29] (also called control constraints or program constraints), which can be imposed onto the controlled system as additional constraint functions. The servo constraints enforce the desired motion along prescribed trajectories and thus can be used to partially prescribe the motion of the discrete mechanical systems. The use of servo constraints makes possible a simulation approach to the inverse dynamics of underactuated multibody systems.

The partial specification of the motion of underactuated multibody systems by means of servo constraints typically leads to a problem formulation in terms of differential-algebraic equations (DAEs). If minimal coordinates are used, the differential part of the DAEs corresponds to the equations of motion (4.1b), whereas the algebraic part is related to the servo constraints (4.1c). In the special case of fully actuated multibody systems, the simulation approach to the inverse dynamics problem yields index-3 DAEs that can be integrated in analogy to the DAEs corresponding to constrained mechanical systems (see, e.g., [89]). However, the situation changes considerably if underactuated mechanical systems are dealt with. In this type of systems, the number of degrees of freedom exceeds the number of controls. The use of servo constraints in the context of underactuated multibody systems leads to a broad diversity of servo constraint problems (see, in particular, [26, 33, 82]). One indicator of problem diversity is the (differentiation) index [6] of the underlying DAEs that typically ranges from three to five and even higher. The index of a set of DAEs is an important characteristic, which is a measure of singularity of the DAEs and indicates difficulty in their numerical treatment. The index of a DAE system denotes also the number of times the algebraic equations of the system need to be differentiated with respect to time to get a set of ordinary differential equations in all the involved variables [30]. Consequently, to facilitate a stable numerical integration, some kind of index reduction approach needs to be applied. This issue will be discussed in detail in the subsequent chapters. The other indicator of problem diversity is related to the differential flatness [78] of underactuated mechanical systems. If the underactuated system is differentially flat, the analytical solution can be obtained through the flatness-based formulation. If the underactuated system is non-flat, the stability of the internal dynamics is of paramount importance and ensures the controllability of the system [26, 33, 82]. Here only differentially flat systems are considered.

4.1.1 Generalized coordinates formulation

Servo constraints can be appended to the previously derived equations of motion to formulate the inverse dynamics control problem of underactuated mechanical systems. Using minimal coordinates, the DAEs governing the inverse dynamics of discrete underactuated mechanical systems consist of the equations of motion and the servo constraints [25, 27]. In particular, the equations of motion have the form

$$\dot{\boldsymbol{\mu}} = \boldsymbol{\nu} \tag{4.1a}$$

$$\boldsymbol{\mathcal{M}}(\boldsymbol{\mu})\dot{\boldsymbol{\nu}} = \boldsymbol{f}(\boldsymbol{\mu}, \dot{\boldsymbol{\mu}}) + \boldsymbol{\mathcal{B}}^T(\boldsymbol{\mu})\boldsymbol{u} \tag{4.1b}$$

with minimal coordinates $\boldsymbol{\mu} \in \mathbb{R}^f$, generalized velocities $\boldsymbol{\nu} \in \mathbb{R}^f$, positive definite mass matrix $\boldsymbol{\mathcal{M}} \in \mathbb{R}^{f,f}$, generalized forces $\boldsymbol{f} \in \mathbb{R}^f$, control inputs $\boldsymbol{u} \in \mathbb{R}^a$, and input transformation matrix $\boldsymbol{\mathcal{B}} \in \mathbb{R}^{a,f}$. Furthermore, $t \in \mathbb{I}$ denotes the time, and

$\mathbb{I} = [t_0, t_f] \subset \mathbb{R}$ is the time interval of interest. The equations of motion are subject to the servo constraints

$$s(\mu) = \gamma(t) \tag{4.1c}$$

where $\gamma(t) : \mathbb{I} \to \mathbb{R}^a$ is the desired output function. Note that the number a of control inputs is assumed to be equal to the number of independent servo constraints. Correspondingly, the Jacobian of the servo constraints $\mathcal{S}(\mu) := Ds(\mu)$ is assumed to have full (row) rank. At this point the control inputs are regarded as variables [63, Ch. 3.6]. The attention is focused on underactuated mechanical systems in which the number of controls is lower than the number of degrees of freedom, that is, $a < f$.

A distinguishing feature of the DAEs (4.1a)–(4.1c) is that, in general, $\mathcal{B} \neq \mathcal{S}$. This is in sharp contrast to mechanical systems subject to holonomic constraints. The difference between holonomic and servo constraints is further reflected in the rank of the matrix $\mathcal{P} := \mathcal{S}\mathcal{M}^{-1}\mathcal{B}^T$ and in the index of the DAEs (4.1a)–(4.1c). A precise definition of the differentiation index, denoted simply by *index*, can be found in [34].

If the matrix $\mathcal{P}$ has full rank (equal to a), then there should exist an invertible matrix $\mathcal{H} \in \mathbb{R}^{a,a}$ such that $\mathcal{B} = \mathcal{H}\mathcal{S}$. This implies that there exist Lagrange multipliers $\lambda \in \mathbb{R}^a$ such that $\mathcal{B}^T u = \mathcal{S}^T \lambda$. Accordingly, the DAEs (4.1a)–(4.1c) assume the well-known structure of the equations of motion pertaining to (holonomically) constrained mechanical systems written in terms of redundant coordinates. In this special case, the DAEs (4.1a)–(4.1c) are known to have the index of 3. Using the terminology introduced by Blajer [25], this case is called the orthogonal realization of the servo constraints.

In general the matrix $\mathcal{P}$ is rank deficient and the realization of the servo constraints is either mixed orthogonal-tangential or purely tangential in the sense of Blajer [25]. Then the so-called controlled and constrained subspaces do not coincide. In particular, the rank of the matrix $\mathcal{P}$ measures the number of directions of the constrained space which can be directly actuated by the control inputs [27]. For $\operatorname{rank}(\mathcal{P}) < a$ the index of the DAEs (4.1a)–(4.1c) always exceeds 3.

It has already been mentioned that many examples of mechanical systems employing servo constraints lead to DAEs of index 5. Nevertheless, there are examples with arbitrarily high index, see Example 2 in [25]. In the present work crane models that typically yield DAEs of index 5 (see, for example [27, 30, 31]) are considered first. Similarly, the motion of more involved crane-type manipulators such as the wire mechanism dealt with in [53] is governed by DAEs of index 5. This wire mechanism example will be discussed in the sequel.

4.1.2 Rotationless formulation

The underactuated multibody systems can be formulated by a specific rotationless formulation [17] in terms of redundant coordinates as well. One of the main features of the rotationless formulation is the constant mass matrix. Using redundant coordinates, the servo constraints can be expressed by time specified outputs, such as the load coordinates of the end-effector. The servo constraints can be easily appended to the DAEs pertaining to the rotationless formulation of multibody dynamics. This yields a mixed set of standard (passive) constraints and servo constraints. The motion of the discrete underactuated mechanical systems under consideration is governed by differential-algebraic equations with a mixed set of holonomic (4.2d) and control constraints (4.2c).

$$\dot{q} - v = \mathbf{0} \tag{4.2a}$$
$$M\dot{v} + \nabla V(q) + G^T(q)\boldsymbol{\lambda} + B^T u = \mathbf{0} \tag{4.2b}$$
$$c(q,t) = \mathbf{0} \tag{4.2c}$$
$$\boldsymbol{\Phi}(q) = \mathbf{0} \tag{4.2d}$$

Here, the vector of redundant coordinates is denoted by $q \in \mathbb{R}^n$ which specifies the configuration of the underactuated mechanical system at time t. The vector of redundant velocities is expressed by $v = \dot{q}$, where a superposed dot indicates differentiation with respect to time. Accordingly, the configuration vector q and the velocity vector v comprise the vector of state space coordinates (q, v) (see, for example, Rosenberg [77]). The mass matrix $M \in \mathbb{R}^{n\times n}$ is assumed to be constant, symmetric and nonsingular. The kinetic energy of the underactuated system can be written as

$$T(v) = \frac{1}{2} v \cdot M v \tag{4.3}$$

Moreover, the potential energy function is denoted by $V(q) \in \mathbb{R}$. The holonomic constraints are represented by a vector of geometric functions $\boldsymbol{\Phi}(q) \in \mathbb{R}^m$ and the associated constraint Jacobian is described by $G(q) = D\boldsymbol{\Phi}(q) \in \mathbb{R}^{m\times n}$ and $\boldsymbol{\lambda} \in \mathbb{R}^m$ represents a vector of Lagrange-multipliers, which specify the relative magnitude of the constraint forces. The m holonomic constraints are assumed to be independent.

Due to the presence of the holonomic constraints, the constrained underactuated mechanical system has $\tilde{n} = n - m$ degrees of freedom. The corresponding configuration space of the underactuated system is given by

$$Q = \{q(t) \in \mathbb{R}^n \mid \Phi(q) = 0\} \tag{4.4}$$

Furthermore, a vector of servo constraint functions $c(q,t) \in \mathbb{R}^{\tilde{m}}$ is expressed in the form of

$$c(q,t) = s(q) - \gamma(t) \tag{4.5}$$

The servo constraints serve the purpose of partially specifying the motion of underactuated systems ($\tilde{m} < \tilde{n}$) and the system outputs are specified by $s(q) \in \mathbb{R}^{\tilde{m}}$ together with the given desired trajectory $\gamma(t) \in \mathbb{R}^{\tilde{m}}$. Note that the servo constraints comprise rheonomic holonomic constraints as well. The formulation of control constraints in Equation (4.5) is much simpler compared to the formulation in Equation (4.1c) and this will be demonstrated in the numerical examples in the sequel. The corresponding actuator forces are determined by the control inputs $u \in \mathbb{R}^{\tilde{m}}$ in conjunction with the input transformation matrix $B \in \mathbb{R}^{\tilde{m} \times n}$.

4.1.3 Reduced formulation of the DAEs

The generalized coordinates formulation in terms of minimal coordinates can also be obtained by applying the null space method introduced in Section 2.3.2 to the DAEs (4.2a)–(4.2d) emanating from the rotationless formulation in terms of redundant coordinates. This yields the same equations of motion in terms of minimal coordinates as the DAEs (4.1a)–(4.1c).

Assume that it is possible to choose $\tilde{n}$ generalized coordinates $\mu \in U \subset \mathbb{R}^{\tilde{n}}$ for the parameterization of the configuration manifold Q. Then there exists a mapping $F : U \mapsto Q$ such that

$$q = F(\mu) \tag{4.6}$$

Admissible velocities $v \in T_q Q = \text{null}(G(q))$ can be written in the form

$$v = P\nu \tag{4.7}$$

with the generalized velocities $\boldsymbol{v} = \dot{\boldsymbol{\mu}}$ and the null space matrix $\boldsymbol{P} = D\boldsymbol{F}(\boldsymbol{\mu})$. Since the columns of $\boldsymbol{P} \in \mathbb{R}^{n \times \tilde{n}}$ span the null space of $\boldsymbol{G} \in \mathbb{R}^{m \times n}$, this implies

$$\boldsymbol{G}\boldsymbol{P} = \boldsymbol{0} \tag{4.8}$$

Using Equation (4.7), the reduced form of the kinetic energy $\widetilde{T}$ is given by

$$\widetilde{T} = \frac{1}{2}\boldsymbol{v} \cdot \widetilde{\boldsymbol{M}}\boldsymbol{v} \tag{4.9}$$

with the reduced mass matrix

$$\widetilde{\boldsymbol{M}} = \boldsymbol{P}^T \boldsymbol{M} \boldsymbol{P} \tag{4.10}$$

Note that the mass matrix $\widetilde{\boldsymbol{M}}$ is generally configuration dependent and assumed to be positive definite. Premultiplying Equation (4.2b) by $\boldsymbol{P}^T$ and making use of Equation (4.7) and (4.8) yield the reduced formulation

$$\dot{\boldsymbol{\mu}} - \boldsymbol{v} = \boldsymbol{0} \tag{4.11a}$$

$$\widetilde{\boldsymbol{M}}\dot{\boldsymbol{v}} + \boldsymbol{P}^T \boldsymbol{M} \dot{\boldsymbol{P}} \boldsymbol{v} + \nabla \widetilde{V}(\boldsymbol{\mu}) + \widetilde{\boldsymbol{B}}^T \boldsymbol{u} = \boldsymbol{0} \tag{4.11b}$$

$$\widetilde{\boldsymbol{c}}(\boldsymbol{\mu}, t) = \boldsymbol{0} \tag{4.11c}$$

where the servo constraints are given by

$$\widetilde{\boldsymbol{c}}(\boldsymbol{\mu}, t) = \widetilde{\boldsymbol{s}}(\boldsymbol{\mu}) - \boldsymbol{\gamma}(t) \tag{4.12}$$

is obtained by inserting Equation (4.6) into the servo constraints (4.5). Furthermore,

$$\nabla \widetilde{V}(\boldsymbol{\mu}) = \boldsymbol{P}^T \nabla V(\boldsymbol{q}) \quad \text{and} \quad \widetilde{\boldsymbol{B}}^T = \boldsymbol{P}^T \boldsymbol{B}^T \tag{4.13}$$

The resulting DAEs (4.11a)−(4.11c) in terms of generalized coordinates can be regarded as the starting point for index reduction approaches, such as the Blajer-type projection method [25, 27] and the newly proposed index reduction by minimal extension method [2].

4.2 Numerical integration of underactuated systems with servo constraints

In underactuated mechanical systems, more challenging problems may arise due to the underactuation property. In this case the Jacobian of the servo constraints does not span the space of the control inputs any more. This case is termed by Blajer and Kołodziejczyk [27] mixed orthogonal-tangent realization of servo constraints. Consequently, the index of the corresponding DAEs exeeds three in general and this makes the numerical integration of underactuated mechanical systems and the simulation of index-5 problems much more demanding. Even the popular Radau IIa scheme, a Runge-Kutta method with three stages, which is a method of order 5 for ODEs, does not converge for general index-5 problems. Thus, index reduction methods are preferred to reduce the index of the DAEs to 3 or even lower. To yield an index reduction from 5 to 3, Blajer and Kołodziejczyk [27] have proposed a specific projection technique that has been further refined in [31]. The projection approach requires the computation of time-dependent [27, 31] or constant Boolean-type [20, 21] projection matrices in order to split the dynamics of the underactuated system into constrained and unconstrained parts.

The aim here is to get a reformulation of the DAEs which is amenable to a direct discretization. To this end, the specific projection method [20, 21, 27, 31, 94] is applied to the underlying DAEs in terms of generalized coordinates and redundant coordinates, respectively.

4.2.1 Projected formulation in terms of generalized coordinates

In the inverse dynamics formulation, the equations of motion in terms of generalized coordinates can be derived in different ways. The set of DAEs (4.1a)−(4.1c) in terms of generalized coordinates is identical to the resulting set of governing equations (4.11a)−(4.11c) derived by applying the null space method to the rotationless formulation. In the following the latter one will be used as the starting point to apply the projection method proposed in [27].

Differentiating the servo constraints (4.11c) twice with respect to time yields the consistency condition (constraint condition at the acceleration level)

$$\frac{d^2}{dt^2}\widetilde{c}(\mu, t) = \widetilde{C}(\mu)\dot{v} + \widetilde{\xi} = \mathbf{0} \tag{4.14}$$

together with the constraint Jacobian

$$\widetilde{C}(\mu) = D\widetilde{s}(\mu) \tag{4.15}$$

and the constraint-induced acceleration [27]

$$\widetilde{\xi} = \dot{\widetilde{C}}v - \ddot{\gamma} \tag{4.16}$$

Note that the initial values of state variables should satisfy the constraint condition

$$\widetilde{s}(\mu_0) - \gamma(t_0) = 0 \tag{4.17}$$

and the constraint condition at the velocity level

$$\widetilde{C}(\mu_0)v_0 - \dot{\gamma}(t_0) = 0 \tag{4.18}$$

To perform the projection method, a suitable projection matrix $\widetilde{D} \in \mathbb{R}^{\tilde{n}\times(\tilde{n}-\tilde{m})}$ needs to be devised, such that the condition

$$\text{rank}(\widetilde{D}) = \tilde{n} - \tilde{m} \tag{4.19}$$

and the relationship

$$\widetilde{C}\widetilde{D} = \mathbf{0} \tag{4.20}$$

are satisfied.

The projections can be accomplished by premultiplying the dynamic equation (4.11b) with $\widetilde{C}\widetilde{M}^{-1}$ and $\widetilde{D}^T$, respectively. Premultiplying Equation (4.11b) by $\widetilde{C}\widetilde{M}^{-1}$ and using the servo constraint condition at the acceleration level (4.14) yields the projection of Equation (4.11b) into the constrained subspace

$$-\widetilde{\xi} + \widetilde{C}\widetilde{M}^{-1}\{P^T M\dot{P}v + \nabla\widetilde{V}(\mu) + \widetilde{B}^T u\} = \mathbf{0} \tag{4.21}$$

which is called the orthogonal projection [27] of Equation (4.11b).

Next, premultiplying Equation (4.11b) by $\widetilde{D}^T$ yields the projection of Equation (4.11b) into the unconstrained subspace

$$\widetilde{D}^T\{\widetilde{M}\dot{v} + P^T M\dot{P}v + \nabla\widetilde{V}(\mu) + \widetilde{B}^T u\} = \mathbf{0} \tag{4.22}$$

which is called the tangential projection [27] of Equation (4.11b). After the projection procedure, Equation (4.11b) can be replaced by Equation (4.21) and (4.22). This provides the projected formulation in terms of generalized coordinates. Then the governing equations of motion can be summarized as

$$\dot{\mu} - \nu = 0 \tag{4.23a}$$

$$\widetilde{D}^T \widetilde{M} \dot{v} + \widetilde{D}^T \{P^T M \dot{P} v + \nabla \widetilde{V}(\mu) + \widetilde{B}^T u\} = 0 \tag{4.23b}$$

$$\widetilde{C} \widetilde{M}^{-1} \{P^T M \dot{P} v + \nabla \widetilde{V}(\mu) + \widetilde{B}^T u\} - \widetilde{\xi} = 0 \tag{4.23c}$$

$$\widetilde{c}(\mu, t) = 0 \tag{4.23d}$$

Similar to the semi-explicit DAEs [6, 21], the resulting set of equations (4.23a)–(4.23d) can be cast into the form

$$\begin{aligned} H(y)\dot{y} &= f(y, z, t) \\ 0 &= h(y, z, t) \end{aligned} \tag{4.24}$$

where, in the present case,

$$y = \begin{bmatrix} \mu \\ \nu \end{bmatrix} \quad \text{and} \quad z = u \tag{4.25}$$

Note that the application of the tangential projection (4.22) yields a reduction of the number of differential equations from $\tilde{n}$ in (4.11b) to $\tilde{n} - \tilde{m}$ in (4.23b) and the orthogonal projection (4.21) yields $\tilde{m}$ algebraic equations in (4.23c). Thus the size-reduction of the differential part is accompanied by an increase of the algebraic equations from $\tilde{m}$ in DAEs (4.11a)–(4.11c) to $2\tilde{m}$ in DAEs (4.23a)–(4.23d). After the application of the projected formulation, the (differentiation) index has been reduced. If the index of the original DAEs (4.11a)–(4.11c) is five, the index of the DAEs (4.23a)–(4.23d) is reduced to three. Then a direct time discretization can be applied to the DAEs (4.23a)–(4.23d) and the numerical solution to the inverse dynamics simulation can be obtained.

4.2.2 Projected formulation in terms of redundant coordinates

Similar to Section 4.2.1, the projected formulation can also be applied to the high index DAEs (4.2a)–(4.2d) in terms of redundant coordinates. The projected formulation has been presented in [20, 21] and refined in [94] later. It is closely related to the projection method applied to the formulation in terms of dependent coordinates in Blajer and Kołodziejczyk [30, 31].

Differentiating the servo constraints (4.2c) twice with respect to time yields the condition at the acceleration level

$$\frac{d^2}{dt^2} c(q, t) = C\dot{v} + (\dot{C}v - \ddot{\gamma}) = 0 \tag{4.26}$$

with the constraint Jacobian

$$C = Ds(q) \tag{4.27}$$

Upon introduction of the constraint-induced acceleration [27]

$$\xi = \dot{C}v - \ddot{\gamma} \tag{4.28}$$

the servo constraint condition (4.26) can be written as

$$C\dot{v} = -\xi \tag{4.29}$$

To perform the projection method, an appropriate matrix $D \in \mathbb{R}^{n\times(n-\tilde{m})}$ needs to be set up such that the relationship

$$CD = 0 \tag{4.30}$$

is satisfied.

The orthogonal projection is performed by premultiplying Equation (4.2b) by CM^{-1} and taking into account Equation (4.29). This leads to $\tilde{m}$ algebraic equations

$$-\xi + CM^{-1}\{\nabla V(q) + G^T(q)\lambda + B^T u\} = 0 \tag{4.31}$$

Note that, for simplicity, it has been tacitly assumed that the mass matrix M is non-singular. The tangential projection is performed by premultiplying Equation (4.2b) by D^T. This leads to $n - \tilde{m}$ differential equations

$$D^T\{M\dot{v} + \nabla V(q) + G^T(q)\lambda + B^T u\} = 0 \tag{4.32}$$

By replacing Equation (4.2b) with (4.31) and (4.32), the projected formulation in terms of redundant coordinates is obtained, and the equations of motion are given by

$$\dot{q} - v = 0 \tag{4.33a}$$
$$D^T M\dot{v} + D^T\{\nabla V(q) + G^T(q)\lambda + B^T u\} = 0 \tag{4.33b}$$
$$CM^{-1}\{\nabla V(q) + G^T(q)\lambda + B^T u\} - \xi = 0 \tag{4.33c}$$
$$c(q,t) = 0 \tag{4.33d}$$
$$\Phi(q) = 0 \tag{4.33e}$$

The index of the resulting set of DAEs (4.33a)–(4.33e) has been reduced to a lower number after the application of the projection method. Then the direct time discretization can be performed by applying the numerical integrators (see Chapter 3). The control strategy can thus be obtained for the trajectory tracking control problem. It is worth mentioning that the rotationless formulation of multibody dynamics for trajectory tracking control problems typically yields projection matrices $\boldsymbol{C}$ and $\boldsymbol{D}$ of Boolean (or binary) type [21]. This feature is highly beneficial to the time discretization of the underlying DAEs (4.33a)–(4.33e), which can be written in the form of Equation (4.24) again, with

$$\boldsymbol{y} = \begin{bmatrix} \boldsymbol{q} \\ \boldsymbol{v} \end{bmatrix} \quad \text{and} \quad \boldsymbol{z} = \begin{bmatrix} \boldsymbol{\lambda} \\ \boldsymbol{u} \end{bmatrix} \tag{4.34}$$

Note that the incorporation of servo constraint-induced acceleration in Equation (4.29) turns $\tilde{m}$ of the original differential equations (4.2b) into algebraic equations (4.31).

4.2.3 Projected formulation in terms of dependent coordinates

Generalized coordinates are called independent (minimal) coordinates in the formulation of cranes in Blajer and Kołodziejczyk [30, 31]. Therein dependent (non-minimal) coordinates can be regarded as redundant coordinates. However, due to the use of rotational parameters in the robot coordinates (see Section 4.4), dependent coordinates do have distinction from redundant coordinates which employ directors (see Subsection 2.4.2).

In the simulation of overhead and rotary cranes, dependent coordinates are often divided into two groups (see [30, 53]), which are the robot coordinates $\boldsymbol{p} \in \mathbb{R}^{n-\tilde{m}}$ and the load coordinates $\boldsymbol{x} \in \mathbb{R}^{\tilde{m}}$. They are related through the geometric constraints (4.2d). Then the coordinates can be expressed by

$$\boldsymbol{q} = \begin{bmatrix} \boldsymbol{p} \\ \boldsymbol{x} \end{bmatrix} \tag{4.35}$$

Using these coordinates, the servo constraints (4.5) are simplified to the following trivial form

$$\boldsymbol{x} = \boldsymbol{\gamma}(t) \tag{4.36}$$

and the original DAEs (4.2a)–(4.2d) can be rewritten as

$$\dot{p} - w = 0 \tag{4.37a}$$

$$M(p)\begin{bmatrix}\dot{w}\\ \ddot{\gamma}\end{bmatrix} + \nabla V(p,\gamma) + G^T(p,\gamma)\lambda + B^T u = 0 \tag{4.37b}$$

$$\Phi(p,\gamma) = 0 \tag{4.37c}$$

in which the vector w denotes the robot velocities and the load coordinates x have been replaced by the time specified output function $\gamma(t)$. Note that due to the use of rotational parameters the mass matrix $M(p)$ here is configuration dependent in the dependent coordinates formulation of overhead and rotary cranes.

The projected formulation in terms of dependent coordinates can be applied, which has also been provided in [30, 31]. Differentiating the holonomic constraints (4.37c) twice with respect to time yields the constraint condition at the acceleration level

$$\frac{d^2}{dt^2}\Phi(p,\gamma) = G(q)\dot{v} + \dot{G}(q)v = 0 \tag{4.38}$$

with the constraint Jacobian $G(q) = D\Phi(q) \in \mathbb{R}^{m\times n}$. Again the associated constraint-induced acceleration is introduced as

$$\xi = \dot{G}(q)v \tag{4.39}$$

and Equation (4.38) can be rewritten as

$$G(q)\dot{v} = -\xi \tag{4.40}$$

Then a suitable projection matrix $D \in \mathbb{R}^{n\times(n-m-\tilde{m})}$ can be devised, such that the relationship

$$AD = 0 \tag{4.41}$$

or in an equivalent form

$$\begin{bmatrix}C\\ G\end{bmatrix} D = \begin{bmatrix}CD\\ GD\end{bmatrix} = 0 \tag{4.42}$$

with the expression

$$A = \begin{bmatrix}C\\ G\end{bmatrix} \tag{4.43}$$

is satisfied and the servo constraint Jacobian $\boldsymbol{C}$ is given by Equation (4.27) with the expression $\boldsymbol{s}(\boldsymbol{q}) = \boldsymbol{x}$. Premultiplying Equation (4.37b) by $\boldsymbol{D}^T$ and taking into account the relationship in Equation (4.42)

$$\boldsymbol{G}\boldsymbol{D} = \boldsymbol{0} \tag{4.44}$$

yield the tangential projection

$$\boldsymbol{D}^T \left\{ \boldsymbol{M}(\boldsymbol{p}) \begin{bmatrix} \dot{\boldsymbol{w}} \\ \ddot{\boldsymbol{\gamma}} \end{bmatrix} + \nabla V(\boldsymbol{q}) + \boldsymbol{B}^T \boldsymbol{u} \right\} = \boldsymbol{0} \tag{4.45}$$

In the redundant coordinates formulation, the servo constraint Jacobian $\boldsymbol{C}$ is Boolean type, thus the servo constraint condition at the acceleration level (see Equation (4.29)) can be written as

$$\boldsymbol{C}\dot{\boldsymbol{v}} = \ddot{\boldsymbol{\gamma}} \tag{4.46}$$

Premultiplying Equation (4.37b) by $\boldsymbol{C}\boldsymbol{M}^{-1}$ and making use of Equation (4.46) yield the projection into the specified subspace $\mathcal{C}$. That is

$$\ddot{\boldsymbol{\gamma}} + \boldsymbol{C}\boldsymbol{M}^{-1}\{\nabla V(\boldsymbol{q}) + \boldsymbol{G}^T \boldsymbol{\lambda}\} = \boldsymbol{0} \tag{4.47}$$

Here, the relationship of the inner product of the specified subspace $\mathcal{C}$ and the controlled subspace $\mathcal{B}$ (for more details, see [30, 31]) is given by

$$\boldsymbol{C}\boldsymbol{M}^{-1}\boldsymbol{B}^T = \boldsymbol{0} \tag{4.48}$$

because the two $\tilde{m}$-subspaces $\mathcal{C}$ and $\mathcal{B}$ are complementary and disjoint.

Premultiplying Equation (4.37b) by $\boldsymbol{G}\boldsymbol{M}^{-1}$ with the incorporation of Equation (4.40) yields the projection into the constrained subspace $\mathcal{G}$. That is

$$-\boldsymbol{\xi} + \boldsymbol{G}\boldsymbol{M}^{-1}\{\nabla V(\boldsymbol{q}) + \boldsymbol{G}^T \boldsymbol{\lambda} + \boldsymbol{B}^T \boldsymbol{u}\} = \boldsymbol{0} \tag{4.49}$$

The constrained subspace $\mathcal{G}$ has a nonzero inner product with both the controlled subspace $\mathcal{B}$ and the specified subspace $\mathcal{C}$.

If the original set of DAEs (4.37a)–(4.37c) has the index of 5, after the application of the projection method, the governing equations are the following $2n - \tilde{m} + m$ index-3 DAEs in the same number of variables $\boldsymbol{p}$, $\boldsymbol{w}$, $\boldsymbol{u}$ and $\boldsymbol{\lambda}$:

$$\dot{p} - w = 0 \tag{4.50a}$$

$$D^T M(p) \begin{bmatrix} \dot{w} \\ \ddot{\gamma} \end{bmatrix} + D^T \{\nabla V(p, \gamma) + B^T u\} = 0 \tag{4.50b}$$

$$\ddot{\gamma} + C M^{-1} \{\nabla V(p, \gamma) + G^T \lambda\} = 0 \tag{4.50c}$$

$$-\xi + G M^{-1} \{\nabla V(p, \gamma) + G^T \lambda + B^T u\} = 0 \tag{4.50d}$$

$$\Phi(p, \gamma) = 0 \tag{4.50e}$$

which can be written in the form of (4.24), with

$$y = \begin{bmatrix} p \\ w \end{bmatrix} \quad \text{and} \quad z = \begin{bmatrix} \lambda \\ u \end{bmatrix} \tag{4.51}$$

It is to be noted that only the case, in which the dimension of the constrained subspace $\mathcal{G}$ is lower than the dimension of the specified space $\mathcal{C}$, is considered, that is, $m < \tilde{m}$.

4.2.4 Numerical discretization

The projected formulations in terms of generalized coordinates (4.23a)–(4.23d), redundant coordinates (4.33a)–(4.33e) and dependent coordinates (4.50a)–(4.50e) yield DAEs in semi-explicit form (4.24). In a first step towards the time discretization of the DAEs a backward Euler-type method is applied. Accordingly, the time-stepping scheme is given by

$$\begin{aligned} H(y_{n+1})(y_{n+1} - y_n) &= \Delta t f(y_{n+1}, z_{n+1}, t_{n+1}) \\ 0 &= h(y_{n+1}, z_{n+1}, t_{n+1}) \end{aligned} \tag{4.52}$$

where Δt is the time step size. The corresponding discretization of the projected formulation in terms of generalized coordinates (4.23a)–(4.23d) leads to the scheme originally proposed by Blajer and Kołodziejczyk [27].

4.3 Differential flatness

Besides the numerical integration, the inverse dynamics control problems of underactuated mechanical systems can be viewed from the perspective of differentially flatness introduced by Fliess et al. [39]. If the underactuated mechanical system is differentially flat, the specified outputs play the role of flat outputs, which can be used to completely determine the motion of the underactuated system. One main property of differential flatness is that all the state variables and control inputs can be directly expressed in terms of the flat outputs and their time derivatives up to a certain order, without integrating any differential equation. The system of DAEs expressed in Equation (4.24) is differentially flat if the following algebraic functions can be obtained:

$$y = f_y(\gamma, \dot{\gamma}, \ldots, \gamma^{(\alpha-1)}) \tag{4.53a}$$

$$u = f_u(\gamma, \dot{\gamma}, \ldots, \gamma^{(\alpha)}) \tag{4.53b}$$

The value of α is by one smaller than the value of index of the DAEs (4.24). More detailed background about differential flatness can be found in [53, 78, 90]. In the following numerical examples, it is shown that differential flatness yields the flatness-based solution, which can be considered as the analytical solution. However, it is generally not feasible to get such an analytical solution for more complicated multibody systems, and then numerical methods such as index reduction approaches are needed to solve the inverse dynamics problems of underactuated multibody systems.

4.4 Numerical examples

The numerical integration approach (projected formulation) and the analytical approach (formulation based on differential flatness) will be applied to two examples, the planar overhead crane and the three-dimensional rotary crane, respectively.

4.4.1 Planar overhead crane

As shown in Fig. 4.1, the planar example of an overhead crane is considered (see also [27, 30]) as a prototypical example of an underactuated mechanical system, which has three degrees of freedom and is composed of the trolley, the winch and the load.

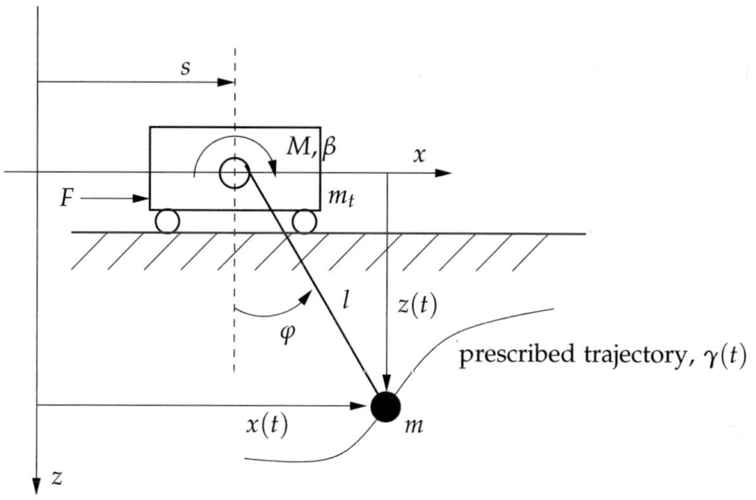

Figure 4.1: The model of the planar overhead crane.

Rotationless formulation in terms of redundant coordinates

A rotationless formulation[1] is first considered, which relies on $n = 4$ redundant coordinates[2] given by

$$q = [s \quad \beta \quad x \quad z]^T \tag{4.54}$$

where the trolley position is specified by s, the rotation angle of the winch is denoted by β, and the load coordinates (see Fig. 4.1) are given by x and z.

Note that the hoisting cable connecting the load with the winch is assumed to be massless and inextensible. The corresponding 4×4 mass matrix is constant and diagonal:

[1] Strictly speaking, the present formulation is not rotationless due to the presence of angle β. However, the present description in terms of redundant coordinates for the planar overhead crane still leads to a constant mass matrix which is a prominent feature of the rotationless formulation. By contrast, the formulation in terms of robot coordinates and load coordinates for the three-dimensional rotary crane leads to configuration-dependent mass matrix which will be shown in the examples later. Therefore, redundant coordinates of this type are also called dependent coordinates by Blajer and Kołodziejczyk. The truly rotationless formulation has been discussed thoroughly in Subsection 2.4 (see also [88]).

[2] Instead of the rotation angle β of the winch, the length l of the cable can also be used.

$$M = \begin{bmatrix} m_t & 0 & 0 & 0 \\ 0 & J & 0 & 0 \\ 0 & 0 & m & 0 \\ 0 & 0 & 0 & m \end{bmatrix} \tag{4.55}$$

in which the trolley mass is m_t, the moment of inertia of the winch is J, and the load mass is m. Furthermore, the radius of the winch is given by r.

Gravity is acting on the system such that

$$V(\boldsymbol{q}) = -mgz \tag{4.56}$$

where the gravitational acceleration is denoted by g. To link the position of the load to the rotation (angle β) of the winch, the holonomic constraint function ($m = 1$)

$$\Phi(\boldsymbol{q}) = (x - s)^2 + z^2 - (r\beta)^2 \tag{4.57}$$

is introduced. Correspondingly, the constraint Jacobian $\boldsymbol{G} = D\Phi(\boldsymbol{q})$ assumes the following form

$$\boldsymbol{G} = 2 \begin{bmatrix} s - x & -r^2\beta & x - s & z \end{bmatrix} \tag{4.58}$$

The specified trajectory of the load is expressed by the servo constraint function given by Equation (4.5) with $\tilde{m} = 2$. In this connection,

$$\boldsymbol{s}(\boldsymbol{q}) = \boldsymbol{C}\boldsymbol{q} = \begin{bmatrix} x \\ z \end{bmatrix} \tag{4.59}$$

with the constant Boolean Jacobian

$$\boldsymbol{C} = D\boldsymbol{s}(\boldsymbol{q}) = \begin{bmatrix} 0 & 0 & 1 & 0 \\ 0 & 0 & 0 & 1 \end{bmatrix} \tag{4.60}$$

Moreover, the prescribed trajectory function $\gamma(t) \in \mathbb{R}^2$ in Equation (4.5) takes the following form

$$\gamma(t) = \begin{bmatrix} x_d(t) \\ z_d(t) \end{bmatrix} \tag{4.61}$$

where the desired time-specified coordinates of the trajectory of the load are $x_d(t)$ and $z_d(t)$, i.e. appropriately smooth functions of time. The actuator forces as control inputs follow from

$$u = \begin{bmatrix} F \\ M \end{bmatrix} \tag{4.62}$$

and the input transformation matrix $\boldsymbol{B}$ is given by

$$\boldsymbol{B} = \begin{bmatrix} -1 & 0 & 0 & 0 \\ 0 & -1 & 0 & 0 \end{bmatrix} \tag{4.63}$$

To summarize, the above quantities completely specify the DAEs (4.2a)−(4.2d) and the resulting set of DAEs has the index of five.

Projected formulation in terms of redundant coordinates

The application of the projected formulation in terms of redundant coordinates (see Subsection 4.2.2) is performed next. It is easy by inspection to find the special Boolean matrix

$$\boldsymbol{D} = \begin{bmatrix} 1 & 0 \\ 0 & 1 \\ 0 & 0 \\ 0 & 0 \end{bmatrix} \tag{4.64}$$

which qualifies as viable projection matrix and satisfies the condition (4.30). Moreover, in view of Equation (4.28) and (4.60), the constraint-induced acceleration is given by

$$\xi = -\ddot{\gamma} \tag{4.65}$$

which is calculated by using Equation (4.61). The prescribed trajectory $\gamma(t)$ can be generated by using a reference function in the motion planning. These quantities are required to set up the DAEs (4.33a)−(4.33e) pertaining to the projected formulation in terms of redundant coordinates. Note that the index of the resulting DAEs (4.33a)−(4.33e) has been reduced to three after the projection method is performed in the planar overhead crane example.

Projected formulation in terms of dependent coordinates

As mentioned above, the present redundant (dependent) coordinates $\boldsymbol{q}$ can be divided into two groups, the robot coordinates $\boldsymbol{p} \in \mathbb{R}^2$ and the load coordinates $\boldsymbol{x} \in \mathbb{R}^2$, which are given by

$$p = \begin{bmatrix} s & \beta \end{bmatrix}^T \quad \text{and} \quad x = \begin{bmatrix} x & z \end{bmatrix}^T \tag{4.66}$$

In this simple example, the matrix A in Equation (4.43) is given by

$$A = \begin{bmatrix} 0 & 0 & 1 & 0 \\ 0 & 0 & 0 & 1 \\ s-x & -r^2\beta & x-s & z \end{bmatrix} \tag{4.67}$$

Then the projection matrix D can be easily found by guess, such as

$$D = \begin{bmatrix} 1 & \frac{s-x}{r^2\beta} & 0 & 0 \end{bmatrix}^T \tag{4.68}$$

which satisfies the condition (4.41). Furthermore, in view of Equation (4.39) and (4.58), the constraint-induced acceleration of constraint (4.57) is calculated by

$$\xi = 2\left[(\dot{s}-\dot{x})\dot{s} - r^2\dot{\beta}^2 + (\dot{x}-\dot{s})\dot{x} + \dot{z}^2\right] \tag{4.69}$$

Then the above quantities can specify the index-3 DAEs (4.50a)–(4.50e) pertaining to the projected formulation in terms of dependent coordinates.

Generalized coordinates formulation in terms of minimal coordinates

To set up the problem formulation of the overhead crane in terms of minimal coordinates, the reduction procedure (see Subsection 4.1.3) on the basis of the rotationless formulation in terms of redundant coordinates is performed to achieve the transition to the DAEs (4.11a)–(4.11c) in terms of minimal coordinates. To this end, the $\tilde{n} = 3$ dimensional configuration manifold is parameterized with minimal coordinates

$$\mu = \begin{bmatrix} s \\ l \\ \varphi \end{bmatrix} \tag{4.70}$$

Here, the variable l denotes the length of the inextensional hoisting cable connecting (the axis of) the winch with the load. Moreover, the variable φ measures the angle between the vertical and the hoisting cable (see Fig. 4.1).

There exists a mapping $\boldsymbol{F}: U \mapsto Q \subset \mathbb{R}^4$, which can be written as

$$\boldsymbol{q} = \boldsymbol{F}(\boldsymbol{\mu}) = \begin{bmatrix} s \\ l/r \\ s + l\sin\varphi \\ l\cos\varphi \end{bmatrix} \tag{4.71}$$

Then the associated Jacobian yields a convenient null space matrix $\boldsymbol{P} = D\boldsymbol{F}(\boldsymbol{\mu})$ of the following form

$$\boldsymbol{P} = \begin{bmatrix} 1 & 0 & 0 \\ 0 & 1/r & 0 \\ 1 & \sin\varphi & l\cos\varphi \\ 0 & \cos\varphi & -l\sin\varphi \end{bmatrix} \tag{4.72}$$

Now, a straightforward calculation yields the additional terms that appear in Equation (4.11a)–(4.11c):

$$\widetilde{\boldsymbol{M}} = \boldsymbol{P}^T\boldsymbol{M}\boldsymbol{P} = \begin{bmatrix} m_t + m & m\sin\varphi & ml\cos\varphi \\ m\sin\varphi & m + J/r^2 & 0 \\ ml\cos\varphi & 0 & ml^2 \end{bmatrix} \tag{4.73a}$$

$$\boldsymbol{P}^T\boldsymbol{M}\dot{\boldsymbol{P}}\boldsymbol{\nu} = \begin{bmatrix} 2m\dot{l}\dot{\varphi}\cos\varphi - ml\dot{\varphi}^2\sin\varphi \\ -ml\dot{\varphi}^2 \\ 2ml\dot{l}\dot{\varphi} \end{bmatrix} \tag{4.73b}$$

$$\nabla\widetilde{V}(\boldsymbol{\mu}) = \boldsymbol{P}^T\nabla V(\boldsymbol{q}) = \begin{bmatrix} 0 \\ -mg\cos\varphi \\ mgl\sin\varphi \end{bmatrix} \tag{4.73c}$$

$$\widetilde{\boldsymbol{B}}^T = \boldsymbol{P}^T\boldsymbol{B}^T = \begin{bmatrix} -1 & 0 \\ 0 & -1/r \\ 0 & 0 \end{bmatrix} \tag{4.73d}$$

$$\boldsymbol{s}(\boldsymbol{\mu}) = \boldsymbol{C}\boldsymbol{F} = \begin{bmatrix} s + l\sin\varphi \\ l\cos\varphi \end{bmatrix} \tag{4.73e}$$

Projected formulation in terms of minimal coordinates

The projected formulation in terms of minimal coordinates (see Subsection 4.2.1) can now be performed. According to Equation (4.15), the constraint Jacobian is given by

$$\widetilde{C} = D\widetilde{s}(\boldsymbol{\mu}) = \boldsymbol{CP} = \begin{bmatrix} 1 & \sin\varphi & l\cos\varphi \\ 0 & \cos\varphi & -l\sin\varphi \end{bmatrix} \tag{4.74}$$

Moreover, in view of Equation (4.16), the constraint-induced acceleration can be obtained. That is

$$\widetilde{\xi} = \begin{bmatrix} 2\,\dot{\varphi}\,\dot{l}\cos\varphi - l\,\dot{\varphi}^2\sin\varphi - \ddot{x}_d \\ -2\,\dot{\varphi}\,\dot{l}\sin\varphi - l\,\dot{\varphi}^2\cos\varphi - \ddot{z}_d \end{bmatrix} \tag{4.75}$$

with the expression

$$\dot{\widetilde{C}}v = \begin{bmatrix} 2\,\dot{\varphi}\,\dot{l}\cos\varphi - l\,\dot{\varphi}^2\sin\varphi \\ -2\,\dot{\varphi}\,\dot{l}\sin\varphi - l\,\dot{\varphi}^2\cos\varphi \end{bmatrix} \tag{4.76}$$

Then a suitable projection matrix can be found by inspection (as in [27])

$$\widetilde{D} = \begin{bmatrix} -1 \\ \sin\varphi \\ \frac{1}{l}\cos\varphi \end{bmatrix} \tag{4.77}$$

which satisfies the complementary condition (4.20). Using the projection matrices, the projection method in terms of minimal coordinates can be applied and the resulting index-3 DAEs (4.23a)–(4.23d) can be set up to perform the inverse dynamics analysis of the planar overhead crane. It is obvious that the quantities are much more complicated than those in the formulation of employing redundant coordiantes.

Analytical solution based on differential flatness

The planar overhead crane can be classified as a differentially flat system, in which the property of differential flatness provides the analytical solution as the reference solution. It will be verified that Equation (4.61) plays indeed the role of flat outputs. This can be easily deduced from the projected formulation in terms of redundant coordinates. In particular, for the overhead crane, Equation (4.33c) yields the following equations

$$2(x_d - s)\lambda + m\ddot{x}_d = 0 \tag{4.78a}$$

$$2z_d\lambda + m(\ddot{z}_d - g) = 0 \tag{4.78b}$$

These equations can be solved for Lagrange multiplier $\lambda(t)$ and the trolley position $s(t)$ in terms of the prescribed outputs together with their derivatives thereof up to the second order:

$$\lambda = \frac{m}{2z_d}(g - \ddot{z}_d) \tag{4.79a}$$

$$s = x_d + \frac{z_d \ddot{x}_d}{g - \ddot{z}_d} \tag{4.79b}$$

Here, the time-specified flat outputs are x_d and z_d. Moreover, the holonomic constraint equation (4.33e) yields the rotation angle $\beta(t)$ of the winch in terms of the prescribed outputs and their derivatives up to the second order:

$$\beta = \pm \frac{z_d}{r(g - \ddot{z}_d)} \sqrt{(\ddot{x}_d)^2 + (g - \ddot{z}_d)^2} \tag{4.80}$$

Eventually, the control inputs (4.62) are determined by Equation (4.33b). Since the product of the transpose of the projection and the transpose of the input transformation matrix yields

$$\boldsymbol{D}^T \boldsymbol{B}^T = -\boldsymbol{I} \tag{4.81}$$

where $\boldsymbol{I}$ is a 2×2 identity matrix, Equation (4.33b) yields

$$\boldsymbol{u} = \boldsymbol{D}^T \{ \boldsymbol{M}\dot{\boldsymbol{v}} + \nabla V(\boldsymbol{q}) + \boldsymbol{G}^T(\boldsymbol{q})\boldsymbol{\lambda} \} \tag{4.82}$$

or in an alternative expression

$$F = m_t \ddot{s} + 2(s - x_d)\lambda \tag{4.83a}$$

$$M = J\ddot{\beta} - 2r^2\beta\lambda \tag{4.83b}$$

Obviously, the control inputs can be expressed by flat outputs along with their time derivatives up to the fourth order, i.e. $\alpha = 4$. It indicates that the index of the original governing DAEs (4.2a)–(4.2d) is five. The above flatness-based analytical solution is provided as the reference solution for the following numerical experiments.

Inverse dynamics simulation

The present numerical experiment is taken from Blajer and Kołodziejczyk [27] and deals with the feedforward control of the planar overhead crane (Fig. 4.1). As flat outputs, the desired trajectory of the load with mass m is prescribed by

$$\gamma(t) = \gamma_0 + (\gamma_f - \gamma_0)c(\tau) \tag{4.84}$$

with the initial position

$$\gamma_0 = \begin{bmatrix} x_d(t_0) & z_d(t_0) \end{bmatrix}^T = \begin{bmatrix} 0 & 4\,\mathrm{m} \end{bmatrix}^T \quad \text{at} \quad t_0 = 0 \tag{4.85}$$

and the final destination

$$\gamma_f = \begin{bmatrix} x_d(t_f) & z_d(t_f) \end{bmatrix}^T = \begin{bmatrix} 5\,\mathrm{m} & 1\,\mathrm{m} \end{bmatrix}^T \quad \text{at} \quad t_f = 3\,\mathrm{s} \tag{4.86}$$

Furthermore, in Equation (4.84), $c(\tau)$ is a 5-6-7-8-9 interpolating polynomial of the following form

$$c(\tau) = 70\tau^9 - 315\tau^8 + 540\tau^7 - 420\tau^6 + 126\tau^5 \tag{4.87}$$

where the term

$$\tau = \frac{t}{t_f - t_0} \tag{4.88}$$

It can be easily checked that Equation (4.84) prescribes a rest-to-rest maneuver of the load. Starting at rest, the initial configuration of the overhead crane can be specified by minimal coordinates

$$\boldsymbol{\mu}_0 = \begin{bmatrix} s_0 & l_0 & \varphi_0 \end{bmatrix}^T = \begin{bmatrix} 0 & 4\,\mathrm{m} & 0 \end{bmatrix}^T \tag{4.89}$$

The remaining parameters are $m_t = 10\,\mathrm{kg}$, $m = 100\,\mathrm{kg}$, $J = 0.1\,\mathrm{kg} \cdot \mathrm{m}^2$ and $r = 0.1\,\mathrm{m}$. For the numerical calculations four different formulations are applied in the following:

GEN: Backward Euler scheme based on DAEs (4.23a)–(4.23d) in terms of generalized coordinates

RED: Backward Euler scheme based on DAEs (4.33a)–(4.33e) in terms of redundant coordinates

ALT: Backward Euler scheme based on DAEs (4.50a)–(4.50e) in terms of dependent coordinates

REF: Reference solution provided by the flatness-based approach

The calculated motion of the inverse dynamics simulation of the overhead crane is illustrated in Fig. 4.2, which contains the snapshots of the system at successive points in time. The simulation yields the numerical results obtained by different projected formulations, which are depicted in Fig. 4.3 and 4.4 for different time step sizes, respectively. It can be observed that all the numerical solutions of the coordinates, the Lagrange multiplier and the control inputs converge to the analytical reference solution if the time step size is reduced.

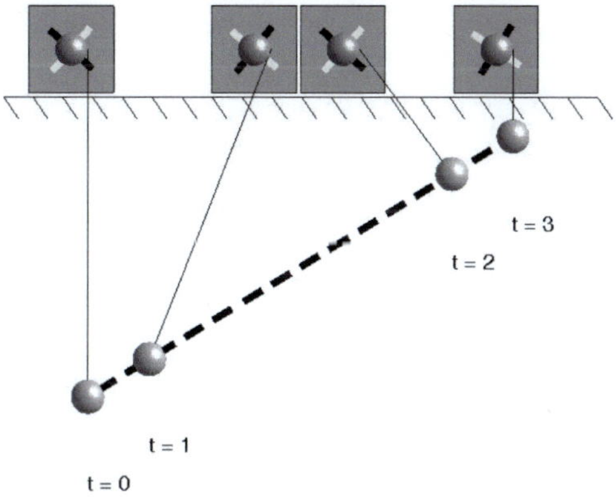

Figure 4.2: Snapshots of the simulation of overhead crane at specific points in time.

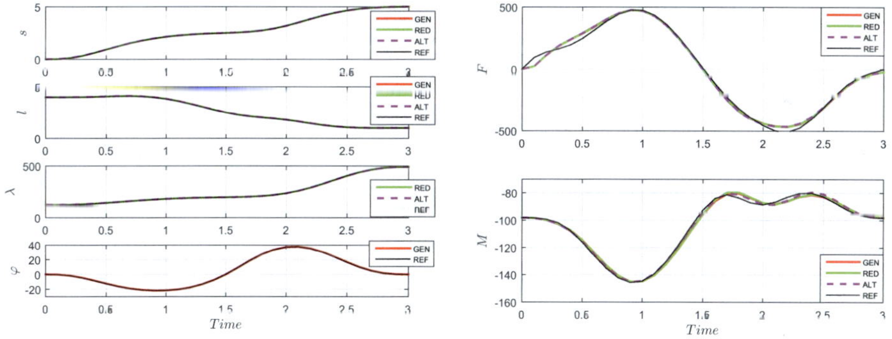

Figure 4.3: Planar overhead crane: Comparison between the numerical results of different projected formulations obtained with $\Delta t = 10^{-1}$ s and the analytical reference solution.

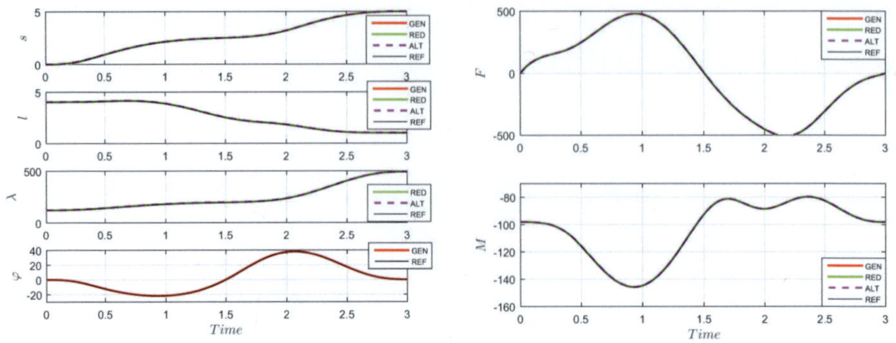

Figure 4.4: Planar overhead crane: Comparison between the numerical results of different projected formulations obtained with $\Delta t = 10^{-3}\,\mathrm{s}$ and the analytical reference solution.

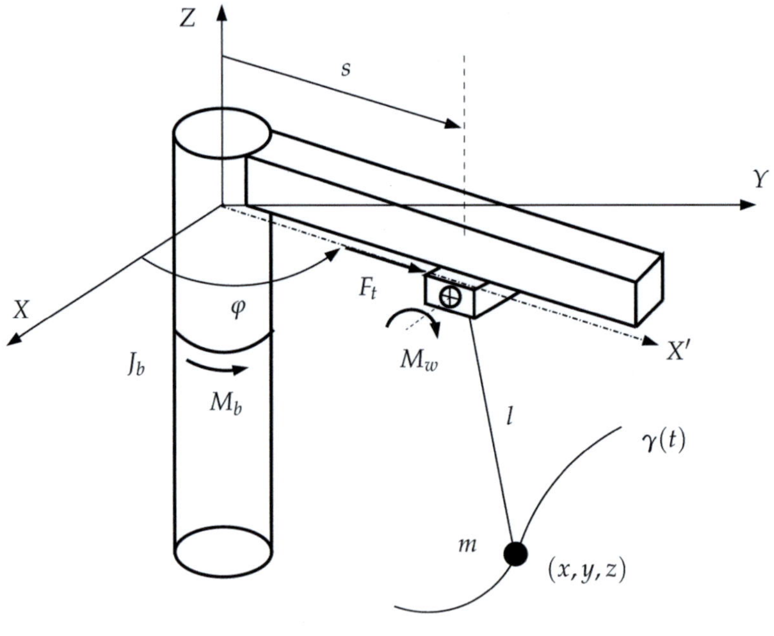

Figure 4.5: The model of the rotary crane.

4.4.2 Three-dimensional rotary crane

The three-dimensional rotary crane, as depicted in Fig. 4.5, is now considered for the inverse dynamics simulation of underactuated systems, which has five degrees of freedom. It has been introduced in the forward dynamics simulation (see Section 2.4).

Rotationless formulation in terms of redundant coordinates

The rotationless formulation in terms of redundant coordinates has been completely presented in the previous three-dimensional rotary crane example (see Section 2.4). The rotationless formulation relies on $n = 42$ redundant coordinates subject to 18 internal constraints (rigidity), 16 external constraints (joints) and 3 additional constraints (coordinate augmentation). Altogether there are $m = 37$ constraints resulting in $\tilde{n} = n - m = 5$ degrees of freedom. Moreover, the configuration vector of redundant coordinates is given by $\boldsymbol{q}$ in Equation (2.103), and the constant mass matrix takes the form of (2.114). The prescribed trajectory function $\gamma(t) \in \mathbb{R}^3$ in Equation (4.5) is given by

$$\gamma(t) = \begin{bmatrix} x_d(t) \\ y_d(t) \\ z_d(t) \end{bmatrix} \tag{4.90}$$

where the coordinates $x_d(t)$, $y_d(t)$ and $z_d(t)$ are the desired time-specified coordinates of the trajectory of the load. The actuator forces (see Fig. 4.5) as control inputs follow from the expression

$$\boldsymbol{u} = \begin{bmatrix} M_b \\ F_t \\ M_w \end{bmatrix} \tag{4.91}$$

in which the torque M_b is acting on the rotary pillar about the vertical rotation axis (Z-axis), the force F_t is acting on the trolley along the girder bridge and the torque M_w is acting on the winch about its rotation axis. The input transformation matrix $\boldsymbol{B}$ is given by

$$\boldsymbol{B} = \begin{bmatrix} \boldsymbol{0}_{1\times 36} & -1 & 0 & 0 \\ \boldsymbol{0}_{1\times 36} & 0 & -1 & 0 \\ \boldsymbol{0}_{1\times 36} & 0 & 0 & -1 \end{bmatrix} \tag{4.92}$$

Accordingly, the actuator forces (control inputs) are incorporated into the present rotationless formulation by applying the coordinate augmentation. All the quantities completely specify the underlying DAEs (4.2a)–(4.2d) which has the index of five.

Projected formulation in terms of redundant coordinates

Similar to the example of the planar overhead crane, the prescribed trajectory of the load leads to $\tilde{m} = 3$ servo constraints of the form (4.5). At this point, the position of the load is determined by

$$s(q) = \varphi^4 = Cq = \begin{bmatrix} x \\ y \\ z \end{bmatrix} \tag{4.93}$$

with the constant Jacobian $C = Ds(q)$ of Boolean type

$$C = \begin{bmatrix} \mathbf{0}_{1\times 36} & 1 & 0 & 0 & \mathbf{0}_{1\times 3} \\ \mathbf{0}_{1\times 36} & 0 & 1 & 0 & \mathbf{0}_{1\times 3} \\ \mathbf{0}_{1\times 36} & 0 & 0 & 1 & \mathbf{0}_{1\times 3} \end{bmatrix} \tag{4.94}$$

To obtain the projected formulation, the projection matrix D satisfying the condition (4.30) can be determined by applying the method in Blajer and Kołodziejczyk [28]. The Jacobian C is assumed to have full row rank (equal to $\tilde{m}$), and it can always be factorized to

$$C = \begin{bmatrix} U & \vdots & W \end{bmatrix} \tag{4.95}$$

so that U has the dimension $\tilde{m} \times (n - \tilde{m})$ and W has the dimension $\tilde{m} \times \tilde{m}$, and the determinant of W is not equal to zero. The orthogonal complement D to C can then be found from

$$D = \begin{bmatrix} I \\ -W^{-1}U \end{bmatrix} \tag{4.96}$$

Then the DAEs (4.33a)–(4.33e) can be set up, which are pertaining to the projected formulation in terms of redundant coordinates.

Projected formulation in terms of dependent coordinates

The dependent coordinates [31] can be employed as redundant coordinates to formulate the problem of the rotary crane as well. Then the robot coordinates $p \in \mathbb{R}^3$ and the load coordinates $x \in \mathbb{R}^3$ are given by

$$p = \begin{bmatrix} \varphi \\ s \\ l \end{bmatrix} \quad \text{and} \quad x = \begin{bmatrix} x \\ y \\ z \end{bmatrix} \tag{4.97}$$

which are related through the passive constraint ($m = 1$)

$$\Phi(p, x) = L - l = \sqrt{(x - s\cos\varphi)^2 + (y - s\sin\varphi)^2 + z^2} - l = 0 \tag{4.98}$$

with the length

$$L = \sqrt{(x - s\cos\varphi)^2 + (y - s\sin\varphi)^2 + z^2} \tag{4.99}$$

Correspondingly, the constraint Jacobian $\boldsymbol{G} = D\Phi(\boldsymbol{p}, \boldsymbol{x})$ assumes the form

$$\boldsymbol{G} = \begin{bmatrix} \dfrac{(x\sin\varphi - y\cos\varphi)s}{L} & \dfrac{s - x\cos\varphi - y\sin\varphi}{L} & -1 & \dfrac{x - s\cos\varphi}{L} & \dfrac{y - s\sin\varphi}{L} & \dfrac{z}{L} \end{bmatrix} \tag{4.100}$$

The specified trajectory of the load is described by the servo constraint function (4.5) with $\tilde{m} = 3$ and that is

$$\boldsymbol{s}(\boldsymbol{q}) = \boldsymbol{C}\boldsymbol{q} = \begin{bmatrix} x \\ y \\ z \end{bmatrix} \tag{4.101}$$

with the constant Boolean Jacobian

$$\boldsymbol{C} = D\boldsymbol{s}(\boldsymbol{q}) = \begin{bmatrix} 0 & 0 & 0 & 1 & 0 & 0 \\ 0 & 0 & 0 & 0 & 1 & 0 \\ 0 & 0 & 0 & 0 & 0 & 1 \end{bmatrix} \tag{4.102}$$

Moreover, the corresponding 6×6 mass matrix is given by

$$\boldsymbol{M}(\boldsymbol{p}) = \begin{bmatrix} J_b + m_t s^2 & 0 & 0 & 0 & 0 & 0 \\ 0 & m_t & 0 & 0 & 0 & 0 \\ 0 & 0 & J_w/r_w^2 & 0 & 0 & 0 \\ 0 & 0 & 0 & m & 0 & 0 \\ 0 & 0 & 0 & 0 & m & 0 \\ 0 & 0 & 0 & 0 & 0 & m \end{bmatrix} \tag{4.103}$$

Here, the inertia value J_b is the sum of the moment of inertia of the girder bridge, the trolley and the winch relative to the axis of rotation, i.e. Z-axis. The inertia value J_w indicates the moment of inertia of the winch relative to its axis of rotation $\boldsymbol{d}_2^3$ (see Fig. 2.9). The mass m_t includes the mass of the trolley and the winch, and m is the mass of the load. Accordingly, there are

$$J_b = J_3^1 + J_3^2 + J_3^3 \quad \text{and} \quad J_w = J_2^3 \tag{4.104}$$

$$m_t = m_2 + m_3 \quad \text{and} \quad m = m_4 \tag{4.105}$$

Note that the mass matrix $M(p)$ is configuration dependent in the formulation in terms of dependent coordinates for the rotary crane example. Furthermore, the generalized and applied forces contained in the force vector f are given by

$$f = \begin{bmatrix} 2m_t s\dot{s}\dot{\varphi} \\ -m_t s\dot{\varphi}^2 \\ 0 \\ 0 \\ 0 \\ -mg \end{bmatrix} \tag{4.106}$$

The control inputs are given by Equation (4.91) and the input transformation matrix is given by

$$B = \begin{bmatrix} -1 & 0 & 0 & 0 & 0 & 0 \\ 0 & -1 & 0 & 0 & 0 & 0 \\ 0 & 0 & -1/r_w & 0 & 0 & 0 \end{bmatrix} \tag{4.107}$$

Considering Equation (4.39) and (4.100), the constraint-induced acceleration ξ can be calculated. The same projection matrix as provided in [31] is used here, that is

$$D^T = \begin{bmatrix} 1 & 0 & \dfrac{(x\sin\varphi - y\cos\varphi)s}{L} & 0 & 0 & 0 \\ 0 & 1 & \dfrac{s - x\cos\varphi - y\sin\varphi}{L} & 0 & 0 & 0 \end{bmatrix} \tag{4.108}$$

Using these quantities, the DAEs (4.50a)–(4.50e) can eventually be set up, which are more tractable to be dealt with numerically. Note that besides the gravitational force the generalized forces in (4.106) should also be considered in the DAEs (4.37a)–(4.37c) for the dependent coordinates formulation of the rotary crane.

Generalized coordinates formulation in terms of minimal coordinates

The $\tilde{n} = 5$ dimensional configuration manifold of the rotary crane can be parameterized with minimal coordinates

$$\mu = \begin{bmatrix} \varphi & s & l & \theta_1 & \theta_2 \end{bmatrix}^T \tag{4.109}$$

where the rotation angle of the girder bridge is given by φ, the trolley position on the girder bridge is s, the length of the hoisting cable is l, and the swing angles are given by θ_1, θ_2 . The governing equations of the system can be derived either by the

Lagrange equations of the second kind (see Subsection 2.4.1) or by the discrete null space method (see Subsection 2.4.2). If the latter method is applied, the mapping $\boldsymbol{F} : U \mapsto Q \in \mathbb{R}^{39}$ can be written as

$$\boldsymbol{q} = \boldsymbol{F}(\boldsymbol{\mu}) = \begin{bmatrix} \boldsymbol{F}^1(\boldsymbol{\mu}) \\ \boldsymbol{F}^2(\boldsymbol{\mu}) \\ \boldsymbol{F}^3(\boldsymbol{\mu}) \\ \boldsymbol{F}^4(\boldsymbol{\mu}) \\ \boldsymbol{F}^5(\boldsymbol{\mu}) \end{bmatrix} \tag{4.110}$$

where

$$\boldsymbol{F}^1(\boldsymbol{\mu}) = \begin{bmatrix} \boldsymbol{\varphi}^1 \\ \cos\varphi \\ \sin\varphi \\ \boldsymbol{e}_3 \cdot \boldsymbol{\varphi}^1 \\ -\sin\varphi \\ \cos\varphi \\ \boldsymbol{e}_3 \cdot \boldsymbol{\varphi}^1 \\ \boldsymbol{e}_3 \end{bmatrix} \tag{4.111}$$

$$\boldsymbol{F}^2(\boldsymbol{\mu}) = \begin{bmatrix} s\cos\varphi \\ s\sin\varphi \\ 0 \\ \cos\varphi \\ \sin\varphi \\ 0 \\ -\sin\varphi \\ \cos\varphi \\ 0 \\ \boldsymbol{e}_3 \end{bmatrix} \tag{4.112}$$

$$F^3(\mu) = \begin{bmatrix} s\cos\varphi \\ s\sin\varphi \\ 0 \\ \cos\varphi\cos\theta \\ \sin\varphi\cos\theta \\ -\sin\theta \\ -\sin\varphi \\ \cos\varphi \\ 0 \\ \cos\varphi\sin\theta \\ \sin\varphi\sin\theta \\ \cos\theta \end{bmatrix} \tag{4.113}$$

$$F^4(\mu) = \begin{bmatrix} (s + l\sin\theta_2)\cos\varphi + l\cos\theta_2\sin\theta_1\sin\varphi \\ (s + l\sin\theta_2)\sin\varphi - l\cos\theta_2\sin\theta_1\cos\varphi \\ -l\cos\theta_2\cos\theta_1 \end{bmatrix} \tag{4.114}$$

$$F^5(\mu) = \begin{bmatrix} \varphi \\ s \\ l/r_w \end{bmatrix} \tag{4.115}$$

Then these mappings can be used to perform the transition from the rotationless formulation to the formulation in terms of minimal coordinates. At this point it is to be noted that the resulting description in terms of minimal coordinates is quite awkward due to the elaborate expressions.

Projected formulation in terms of minimal coordinates

To perform the projected formulation, the constraint Jacobian needs to be calculated. That is,

$$\widetilde{C} = D\widetilde{s}(\mu) = DF^4(\mu) \tag{4.116}$$

and the constraint-induced acceleration is given by

$$\widetilde{\xi} = \dot{\widetilde{C}}v - \ddot{\gamma} \tag{4.117}$$

Similar to the overhead crane example, a suitable projection matrix $\widetilde{D}$ can be computed by symbolic manipulations, which satisfies the relationship (4.20). Besides, it can be computed by Equation (4.96). Note that the projection matrices are cumbersome and much more complicated in the minimal coordinates formulation for the rotary crane example.

Analytical solution based on differential flatness

The rotary crane example can be classified as a differentially flat system. Then the analytical solution [31] can be obtained through purely algebraic manipulations instead of integrating the DAEs (4.37a)–(4.37c). In the rotary crane example, the projection of the dynamic equations in the specified subspace yields three algebraic equations

$$\ddot{x}_d + \frac{x_d - s\cos\varphi}{m_4 L}\lambda = 0 \tag{4.118a}$$

$$\ddot{y}_d + \frac{y_d - s\sin\varphi}{m_4 L}\lambda = 0 \tag{4.118b}$$

$$\ddot{z}_d + g + \frac{z_d}{m_4 L}\lambda = 0 \tag{4.118c}$$

with the length

$$L^2 = (x_d - s\cos\varphi)^2 + (y_d - s\sin\varphi)^2 + z_d^2 \tag{4.119}$$

From the above three nonlinear equations (4.118a)–(4.118c), the variables can be obtained after some manipulations, which are expressed by

$$\lambda = m_4\sqrt{\ddot{x}_d^2 + \ddot{y}_d^2 + (\ddot{z}_d + g)^2} \tag{4.120a}$$

$$s - \sqrt{{A_x}^2 + {A_y}^2} \tag{4.120b}$$

$$\varphi = \arctan\frac{A_y}{A_x} \tag{4.120c}$$

with the expressions

$$A_x = x_d - \frac{z_d\,\ddot{x}_d}{\ddot{z}_d + g} \tag{4.121a}$$

$$A_y = y_d - \frac{z_d\,\ddot{y}_d}{\ddot{z}_d + g} \tag{4.121b}$$

From the constraint equation (4.98), the length of the cable is calculated by

$$l = -\frac{z_d}{\ddot{z}_d + g}\sqrt{\ddot{x}_d^2 + \ddot{y}_d^2 + (\ddot{z}_d + g)^2} \tag{4.122}$$

Then the actuator (control) forces can be determined from Equation (4.50b) in terms of the flat outputs and their time derivatives up to the fourth order, i.e. $\alpha = 4$. This implies that the initial governing DAEs have the index of five. The above analytical solution provides the flatness-based solution to the differentially flat rotary crane for the inverse dynamics analysis. It is shown that all the state variables and control inputs can be algebraically expressed in terms of the desired outputs and their time derivatives up to a certain order. It is obviously seen that the flatness-based solution is featured by enormous complexity, especially for the acceleration $\dot{w}$ and control inputs u, and thus they may be considered as impractical in applications. In contrast, the numerical approach is much more straightforward and applicable.

Inverse dynamics simulation

body	m [kg]	J_1 [kg · m^2]	J_2	J_3	length [m]	width	depth
1	100	216.67	216.67	16.67	5	1	1
2	50	2.08	2.08	2.08	0.5	0.5	0.5
3	3	0.26	0.02	0.26	1	0.2	0.2
4	10	—	—	—	—	—	—

Table 4.1: The data of mass, moment of inertia and dimension of each body of the rotary crane.

The simulation data used for the example of the rotary crane is summarized in Table 4.1. The required trajectory of the load is prescribed by

$$\gamma(t) = \gamma_0 + (\gamma_f - \gamma_0)c(t) \tag{4.123}$$

with the start position

$$\gamma_0 = \begin{bmatrix} 5\,\mathrm{m} & 0 & -5\,\mathrm{m} \end{bmatrix}^T \quad \text{at} \quad t_0 = 0 \tag{4.124}$$

and the target position

$$\gamma_f = \begin{bmatrix} -2\,\mathrm{m} & 2\,\mathrm{m} & -2\,\mathrm{m} \end{bmatrix}^T \quad \text{at} \quad t_f = 20\,\mathrm{s} \tag{4.125}$$

The same reference function $c(t)$ as in [28] is used here, which prescribes a rest-to-rest motion (see Fig. 4.6) of the load and is composed of three phases: the acceleration phase (I) for $0 \leqslant t < 5\,\mathrm{s}$,

$$c_I(t) = \frac{1}{\tau - \tau_0}\left(-\frac{5t^8}{2\tau_0{}^7} + \frac{10t^7}{\tau_0{}^6} - \frac{14t^6}{2\tau_0{}^5} + \frac{7t^5}{2\tau_0{}^4}\right) \tag{4.126}$$

the steady velocity phase (II) for $5\,\mathrm{s} \leqslant t < 15\,\mathrm{s}$,

$$c_{II}(t) = \frac{1}{\tau - \tau_0}\left(t - \frac{\tau_0}{2}\right) \tag{4.127}$$

and the deceleration phase (III) for $15\,\mathrm{s} \leqslant t \leqslant 20\,\mathrm{s}$,

$$c_{III}(t) = 1 + \frac{1}{\tau - \tau_0}\left(-\frac{5(\tau - t)^8}{2\tau_0{}^7} + \frac{10(\tau - t)^7}{\tau_0{}^6} - \frac{14(\tau - t)^6}{2\tau_0{}^5} + \frac{7(\tau - t)^5}{2\tau_0{}^4}\right) \tag{4.128}$$

where $\tau = t_f - t_0$, and τ_0 is the acceleration/deceleration time. Here, $\tau = 20\,\mathrm{s}$, $\tau_0 = 5\,\mathrm{s}$. The reference function $c(t)$ and its time derivatives are illustrated in Fig. 4.6.

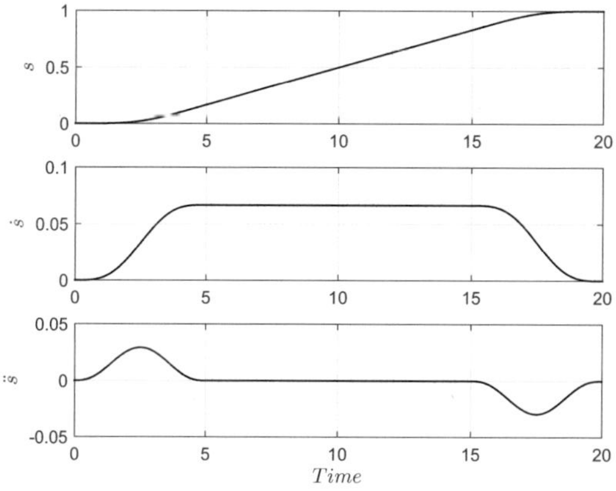

Figure 4.6: Reference function $s(t)$ and its derivatives for the load position.

The design of the reference function $c(t)$ can follow the idea posed in [5]. The synchronized time function (4.123) for the reference load coordinates yields a straight line trajectory from the start position to the target position.

Starting at rest, the initial configuration of the rotary crane is specified by minimal coordinates

$$\boldsymbol{\mu}_0 = \begin{bmatrix}\varphi_0 & s_0 & l_0 & \theta_{1_0} & \theta_{2_0}\end{bmatrix}^T = \begin{bmatrix}0 & 5\,\mathrm{m} & 5\,\mathrm{m} & 0 & 0\end{bmatrix}^T \tag{4.129}$$

The following four formulations are applied to the numerical experiments:

GEN: Backward Euler scheme based on DAEs (4.23a)–(4.23d) in terms of minimal coordinates

RED: Backward Euler scheme based on DAEs (4.33a)–(4.33e) in terms of redundant coordinates

ALT: Backward Euler scheme based on DAEs (4.50a)–(4.50e) in terms of dependent coordinates

REF: Reference solution provided by the flatness-based approach

The calculated motion of the inverse dynamics simulation of the rotary crane is illustrated in Fig. 4.7, which includes the snapshots of the system at successive points in time. Furthermore, the numerical results are obtained by different projected formulations and presented in Fig. 4.8 and Fig. 4.9 for different time step sizes. It can be concluded that the numerical solutions of the coordinates and the control forces converge to the analytical reference solution as the time step size is reduced. Note that the small discrepancy between the projected formulation **ALT** and the other formulations is a value of 0.1, even for very small time step sizes. The distinction may come from the formulation (**ALT**) of the rotary crane.

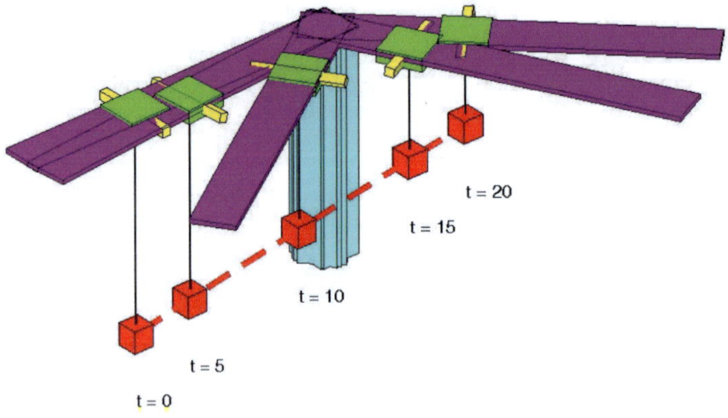

Figure 4.7: Snapshots of the simulation of rotary crane at specific points in time.

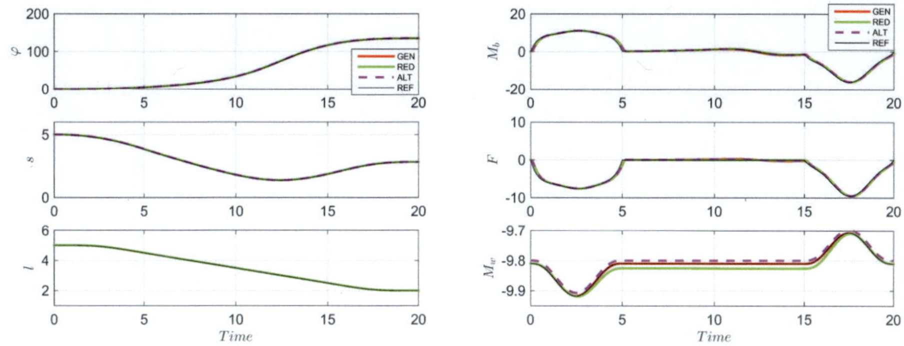

Figure 4.8: Rotary crane: Comparison between numerical results of different formulations obtained with $\Delta t = 10^{-1}$ s and the analytical reference solution.

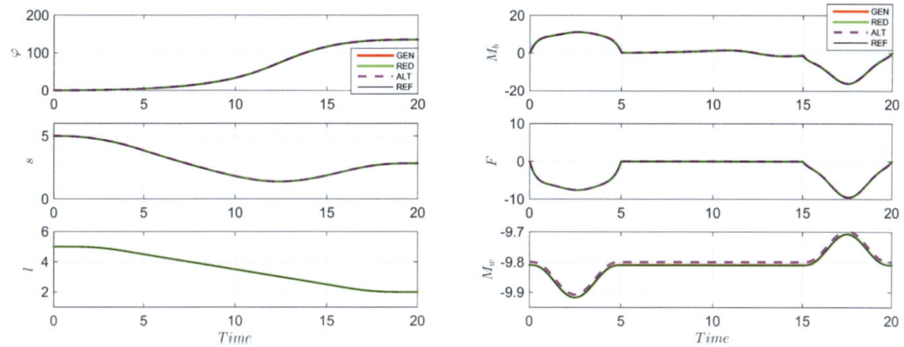

Figure 4.9: Rotary crane: Comparison between numerical results of different formulations obtained with $\Delta t = 5 \times 10^{-3}$ s and the analytical reference solution.

5 Index reduction by minimal extension for the inverse dynamics simulation

So far index reduction by minimal extension [62] has been succesfully applied to circuit simulation [62, Sec. 4], multibody systems [62, Sec. 5] and infinite dimensional systems arising in elastodynamics and flexible multibody systems [1]. However, it has not been applied to underactuated servo constraint problems. To develop the new method for such problems, a class of differentially flat cranes will be considered at first, in which the load coordinates play the role of flat outputs (see also Chapter 4). For the problem formulation, both redundant coordinates and minimal coordinates can be employed within the newly developed method. The formulation of some cranes fits into the more general framework presented in [58].

The underactuated servo constraint problem is governed by differential-algebraic equations with high index (e.g. index 5), which makes the simulation of the problem highly challenging. Therefore, index reduction methods need to be applied to reduce the index of the DAEs to facilitate a stable numerical integration. In Chapter 4 the specific projection method has been used to yield a reduction of the index from 5 to 3. Now an alternative method, which relies on the index reduction by minimal extension originally developed by Kunkel and Mehrmann [62] for more general DAEs, is newly proposed to reduce the index of the DAEs. For the purpose of index reduction, the technique of minimal extension turns out to be especially attractive due to the semi-explicit structure of the DAEs (e.g. see (4.1a)−(4.1b)) and it is not necessary to construct projection matrices as in the projection method. Thus, index reduction by minimal extension can be easily applied to underactuated systems (i.e. $a < f$, e.g. cranes and flexible multibody systems) to reduce the index of the DAEs to 3 or even to 1. As a result, a set of index-3 or index-1 DAEs can be obtained and easily discretized in the numerical integration. Moreover, the reduced index-1 DAEs is purely algebraic and reflects the fact that the system at hand can be classified as differentially flat system. In the end, the DAEs can be solved to determine the associated control inputs, which are required to steer the system such that the prescribed trajectories are tracked. In this way a simulation approach to the feedforward control of multibody systems can be realized. The application of the proposed method will be demonstrated with three representative numerical examples in the following.

5.1 Index reduction by minimal extension

In this section a short introduction to the index reduction approach by minimal extension will be provided. Then its application to the servo constraint problem will be presented.

A common approach for the reduction of the index of general nonlinear DAEs

$$\boldsymbol{F}(t, \boldsymbol{y}, \dot{\boldsymbol{y}}) = \boldsymbol{0} \tag{5.1a}$$

$$\boldsymbol{y}(t_0) = \boldsymbol{y}_0 \tag{5.1b}$$

is given by the derivative array approach [63, Chap. 6.2]. In this equation, $\boldsymbol{y}_0 \in \mathbb{R}^n$ are prescribed initial conditions and $\boldsymbol{F} : \mathbb{I} \times \mathbb{R}^n \times \mathbb{R}^n \to \mathbb{R}^n$. Let the DAEs be of index[1] μ_d. Then all equations need to be differentiated $(\mu_d - 1)$ times and suitable projections are computed to find algebraic and differential equations ,which together form an equivalent system of index 1. It is worth mentioning that a general index concept has been introduced, the so-called strangeness index μ, which generalizes other index concepts, e.g. the concept of the differentiation index μ_d [35]. Further details can be found in [63]. For large systems of high index, the derivative array may become very large and cause memory problems. In addition, high computational effort needs to be invested to find the mentioned projection matrices, which leads to high computational complexity and makes the general method impracticable for large scale problems.

The complexity of the index reduction method can be significantly reduced if additional information about the structure of the system is available, such as DAEs arising in the simulation of multibody systems. This is the case for the semi-explicit DAEs of interest for which the algebraic constraints are explicitly given. The main idea relies

[1] Index indicates the differentiation index in this chapter.

on this structural information about the equations that lead to high index. Hence, it is sufficient to add the derivatives of those equations. This extra information is used to create a reduced size derivative array, so that the computational effort is highly reduced. Even with these improvements, for the reduced size derivative array, local nullspace computations still require large memory storage and arithmetic complexity. To deal with this difficulty, another index reduction concept introduced in [73] is modified. The basic idea of this approach is to introduce new variables, so-called dummy derivatives, to reduce the index. In fact, after introducing so-called dummy variables, projection matrices are even not needed any more. This procedure is then called minimal extension [62].

5.1.1 Minimal extension for mechanical systems

The index reduction technique of minimal extension is applied to the system of equations typically governing the motion of a multibody system (see also [62]). To this end, the DAEs[2]

$$\boldsymbol{M}(\boldsymbol{q})\ddot{\boldsymbol{q}} = \boldsymbol{f}(\boldsymbol{q},\dot{\boldsymbol{q}}) - \boldsymbol{G}^T(\boldsymbol{q})\boldsymbol{\lambda} \tag{5.2a}$$

$$\boldsymbol{\Phi}(\boldsymbol{q}) = \boldsymbol{0} \tag{5.2b}$$

are considered. The redundant coordinates $\boldsymbol{q} \in \mathbb{R}^n$ are subject to m holonomic constraints with associated constraint functions $\boldsymbol{\Phi}(\boldsymbol{q}) \in \mathbb{R}^m$. Lagrange multipliers are given by $\boldsymbol{\lambda} \in \mathbb{R}^m$ and the constraint Jacobian is calculated by $\boldsymbol{G}(\boldsymbol{q}) = D\boldsymbol{\Phi}(\boldsymbol{q}) \in \mathbb{R}^{m,n}$, which is assumed to have full rank. Moreover, $\boldsymbol{M}(\boldsymbol{q}) \in \mathbb{R}^{n,n}$ is a symmetric mass matrix, and $\boldsymbol{f} \in \mathbb{R}^n$ contains the conjugate forces acting on the system, except for the forces of constraint.

It is well known that the present DAEs have index 3 or strangeness index 2 [63, Ex. 4.22]. Since $\boldsymbol{G}(\boldsymbol{q})$ has full rank, there exists an orthogonal matrix $\boldsymbol{Q} \in \mathbb{R}^{n,n}$ such that $\boldsymbol{G}(\boldsymbol{q})\boldsymbol{Q}$ has the block structure

$$\boldsymbol{G}(\boldsymbol{q})\boldsymbol{Q} = \begin{bmatrix} \boldsymbol{G}_1 & \boldsymbol{G}_2 \end{bmatrix} \tag{5.3}$$

[2] For convenience, the equations of motion will not be written in the form of first order.

with an invertible matrix $G_2 \in \mathbb{R}^{m,m}$. The matrix Q then allows to partition the position variables q into $q_1 \in \mathbb{R}^{n-m}$ and $q_2 \in \mathbb{R}^m$ by

$$\begin{bmatrix} q_1 \\ q_2 \end{bmatrix} := Q^T q \tag{5.4}$$

For the reduced derivative array, the two derivatives of the constraints are added to the original system, that is,

$$0 = G(q)\dot{q} \tag{5.5a}$$
$$0 = \dot{G}(q)\dot{q} + G(q)\ddot{q} \tag{5.5b}$$

To avoid the expensive search for projectors, two dummy variables are introduced as

$$\widehat{q}_2 := \dot{q}_2 \tag{5.6a}$$
$$\widetilde{q}_2 := \ddot{q}_2 \tag{5.6b}$$

With the variables q_1, q_2, $\widehat{q}_2$, $\widetilde{q}_2$, and λ, the minimally extended strangeness free (strangeness index 0) system is obtained, which is square. Replacing every occurrence of $\dot{q}_2$ and $\ddot{q}_2$ by the corresponding dummy variables $\widehat{q}_2$ and $\widetilde{q}_2$, the overall system now reads

$$M(q)Q \begin{bmatrix} \ddot{q}_1 \\ \widetilde{q}_2 \end{bmatrix} = f(q_1, q_2, \dot{q}_1, \widehat{q}_2) - G^T(q)\lambda \tag{5.7a}$$
$$0 = \Phi(q_1, q_2) \tag{5.7b}$$
$$0 = G(q)Q \begin{bmatrix} \dot{q}_1 \\ \widehat{q}_2 \end{bmatrix} \tag{5.7c}$$
$$0 = \dot{G}(q)Q \begin{bmatrix} \dot{q}_1 \\ \widehat{q}_2 \end{bmatrix} + G(q)Q \begin{bmatrix} \ddot{q}_1 \\ \widetilde{q}_2 \end{bmatrix} \tag{5.7d}$$

Note that, to prevent clumsy notation, $G(q)$ is used instead of $G(q_1, q_2)$ and similarly $M(q)$ instead of $M(q_1, q_2)$. The following theorem [63, Th. 6.12] shows that the extended system (5.7a)–(5.7d) is strangeness free.

Theorem 5.1. *Consider a multibody system of the form* (5.2a)–(5.2b) *with* $M(q)$ *symmetric and positive definite and suppose that* $G(q)$ *has full row rank. Then the extended system* (5.7a)–(5.7d) *is strangeness free.*

Proof. Since G_2 in (5.3) is square nonsingular, Equation (5.7b) can be solved by means of the implicit function theorem for q_2 in terms of q_1 and Equation (5.7c) can be solved for $\dot{q}_2$ in terms of q_1 and $\dot{q}_1$. Since $M(q)$ is symmetric and positive definite, Equation (5.7a) can be solved for $\ddot{q}_1$ and $\widetilde{q}_2$. Moreover, it follows that

$$W(q) = G(q)M^{-1}G^T(q) \tag{5.8}$$

due to the full row rank of $G(q)$. Hence, $\ddot{q}_1$ and $\widetilde{q}_2$ can be eliminated from Equation (5.7d) and λ can be obtained according to

$$\lambda = W^{-1}(q)\left(\dot{G}(q)Q\begin{bmatrix}\dot{q}_1\\ \widehat{q}_2\end{bmatrix} + G(q)M^{-1}f(q_1,q_2,\dot{q}_1,\widehat{q}_2)\right) \tag{5.9}$$

In the end an ordinary differential equation is obtained in the unknowns q_1 and $\dot{q}_1$. The system has strangeness index $\mu = 0$. □

The proof that the resulting DAEs have index 1 (strangeness index 0) is given in [63, Th. 6.12]. Note that the size of the system has been increased by twice the number of constraints. Thus, for most applications, the system is still of moderate format.

Remark 1. *In general the transformation matrix Q can be found by a Gaussian elimination. In many applications, however, it is possible to guess a permutation matrix Q that yields the needed regular block G_2. In this case, it is possible to choose Q as the identity matrix if a suitable reordering of the variables is assumed and all variables keep their physical meaning.*

5.1.2 Application to the inverse dynamics simulation of cranes

The DAEs describing mechanical systems subject to servo constraints (see Subsection 4.1.1) or, more generally, systems subject to both servo and holonomic constraints (see Subsection 4.1.2), exhibit a semi-explicit structure and are thus very similar to the system dealt with in the previous subsection. Consequently, a similar procedure can be applied to achieve an index reduction for this kind of problems.

In this chapter the focus is placed on the minimal extension approach for cranes. At first the specific redundant (dependent) coordinates are used for the description of the inverse dynamics problem. In addition, the method can also be applied to the corresponding crane formulation in terms of minimal coordinates. This will be shown subsequently in Subsection 5.1.4.

As has been demonstrated in previous works dealing with the description of crane models (see, e.g., Fliess et al. [39, Section 4.1] for a planar overhead crane, Blajer and Kołodziejczyk [31] for a three-dimensional rotary crane, and Heyden and Woernle [53] for a parallel wire-suspended mechanism (see also Section 6.3)), it is especially convenient to divide the crane system into two separate subsystems (Subection 4.4.1). The first subsystem belongs to the motor drives, whereas the second subsystem belongs to the load. Correspondingly, the coordinates are distinguished between *crane (actuated, robot) coordinates* $\boldsymbol{p} \in R^{n-a}$ and *load coordinates* $\boldsymbol{x} \in \mathbb{R}^a$. Using these coordinates, the index-5 DAEs, which govern the controlled motion of cranes, can be written in the form

$$\begin{bmatrix} \boldsymbol{M}_1(\boldsymbol{p}) & \boldsymbol{0} \\ \boldsymbol{0} & \boldsymbol{M}_2 \end{bmatrix} \begin{bmatrix} \ddot{\boldsymbol{p}} \\ \ddot{\boldsymbol{x}} \end{bmatrix} = \begin{bmatrix} \boldsymbol{f}_1(\boldsymbol{p},\dot{\boldsymbol{p}}) \\ \boldsymbol{f}_2(\boldsymbol{x},\dot{\boldsymbol{x}}) \end{bmatrix} + \begin{bmatrix} \boldsymbol{B}_1^T(\boldsymbol{p}) \\ \boldsymbol{0} \end{bmatrix} \boldsymbol{u} - \begin{bmatrix} \boldsymbol{G}_1^T(\boldsymbol{p},\boldsymbol{x}) \\ \boldsymbol{G}_2^T(\boldsymbol{p},\boldsymbol{x}) \end{bmatrix} \boldsymbol{\lambda} \tag{5.10a}$$

$$\boldsymbol{0} = \boldsymbol{\Phi}(\boldsymbol{p},\boldsymbol{x}) \tag{5.10b}$$

$$\boldsymbol{x} = \boldsymbol{\gamma} \tag{5.10c}$$

Here, the first row block in Equation (5.10a) corresponds to the actuated subsystem, whereas the second row block in Equation (5.10a) corresponds to the load. The redundant coordinates

$$\boldsymbol{q} = \begin{bmatrix} \boldsymbol{p} \\ \boldsymbol{x} \end{bmatrix} \tag{5.11}$$

are subject to the holonomic constraints (5.10b) with associated constraint functions $\boldsymbol{\Phi} \in \mathbb{R}^m$ and constraint Jacobian $\boldsymbol{G} = [\boldsymbol{G}_1\ \boldsymbol{G}_2] \in \mathbb{R}^{m,n}$. In this connection,

$$\boldsymbol{G}_1 = \partial_{\boldsymbol{p}} \boldsymbol{\Phi}(\boldsymbol{p},\boldsymbol{x}) \in \mathbb{R}^{m,n-a} \tag{5.12}$$

denotes the partial derivative w.r.t. the crane coordinates $\boldsymbol{p}$, and

$$\boldsymbol{G}_2 = \partial_{\boldsymbol{x}} \boldsymbol{\Phi}(\boldsymbol{p},\boldsymbol{x}) \in \mathbb{R}^{m,a} \tag{5.13}$$

denotes the partial derivative w.r.t. the load coordinates $\boldsymbol{x}$. The holonomic constraints link both subsystems at hand and lead to constraint forces with associated Lagrange multipliers $\boldsymbol{\lambda} \in \mathbb{R}^m$ in Equation (5.10a).

The servo constraints (5.10c) specify the desired trajectory of the load via the prescribed function $\boldsymbol{\gamma} : \mathbb{I} \to \mathbb{R}^a$. The control inputs $\boldsymbol{u} \in \mathbb{R}^a$ regulate the control forces acting on the first subsystem. In this connection, $\boldsymbol{B}_1 \in \mathbb{R}^{a,n-a}$ denotes the input transformation matrix. Besides the constraint and control forces, additional forces acting on the system are contained in the conjugate force vectors $f_1 \in \mathbb{R}^{n-a}$ and $f_2 \in \mathbb{R}^a$. Similarly, the mass matrix is split into the submatrices $M_1 \in \mathbb{R}^{n-a,n-a}$ and $M_2 \in \mathbb{R}^{a,a}$.

Next, the index reduction technique outlined in Subsection 5.1.1 will be applied to the index-5 DAEs (5.10a)–(5.10c) in order to obtain an extended but equivalent system of index 3. Since the holonomic constraint (5.10b) only causes an index of 3, just the derivatives of the servo constraints (5.10c) need to be added. The addition of these two derivatives and the introduction of two dummy variables $\widehat{x} := \dot{x}$ and $\widetilde{x} := \ddot{x}$ lead to the system

$$\begin{bmatrix} M_1(p) & 0 \\ 0 & M_2 \end{bmatrix} \begin{bmatrix} \ddot{p} \\ \widetilde{x} \end{bmatrix} = \begin{bmatrix} f_1(p,\dot{p}) \\ f_2(x,\widehat{x}) \end{bmatrix} + \begin{bmatrix} B_1^T(p) \\ 0 \end{bmatrix} u - \begin{bmatrix} G_1^T(p,x) \\ G_2^T(p,x) \end{bmatrix} \lambda \tag{5.14a}$$

$$0 = \Phi(p,x) \tag{5.14b}$$

$$x = \gamma \tag{5.14c}$$

$$\widehat{x} = \dot{\gamma} \tag{5.14d}$$

$$\widetilde{x} = \ddot{\gamma} \tag{5.14e}$$

As it will be shown in Proposition 1, under certain assumptions, this system of equations has index three.

Proof of index 3

In the following, several assumptions, that are typically satisfied for crane models, will be stated. In particular, these assumptions hold for the examples investigated in Subsection 5.3.1 and 5.3.2. It is emphasized that the most general case will not be analyzed. The analysis is only restricted to the model which ensures that the underlying DAEs (5.10a)–(5.10c) have index 5. The assumptions serve the purpose to minimize technical issues in the subsequent analysis. Furthermore, the assumptions guarantee that the procedure of minimal extension can be applied twice in order to obtain an equivalent system of index 1. This will be shown in Subsection 5.1.3.

Assumption 1. *Consider system* (5.10a)–(5.10c) *with $m \leqslant a \leqslant n-a$. Let $M_1 \in \mathbb{R}^{n-a,n-a}$ be positive definite, and $G_2 \in \mathbb{R}^{m,a}$ have full rank. This implies that there exists a matrix $P_2 \in \mathbb{R}^{a,a-m}$, whose columns span the null space of G_2. Thus*

$$G_2(p,\gamma)P_2(p,\gamma) = 0 \tag{5.15}$$

Define $\overline{z} \in \mathbb{R}^{a-m}$ *by*

$$\overline{z}(t,\boldsymbol{p}) := \boldsymbol{P}_2^T(\boldsymbol{p},\boldsymbol{\gamma})\,(\boldsymbol{f}_2(\boldsymbol{\gamma},\dot{\boldsymbol{\gamma}}) - \boldsymbol{M}_2\ddot{\boldsymbol{\gamma}}) \tag{5.16}$$

and introduce $\overline{\boldsymbol{h}} \in \mathbb{R}^a$ *by*

$$\overline{\boldsymbol{h}}(t,\boldsymbol{p}) := \begin{bmatrix} \boldsymbol{\Phi}(\boldsymbol{p},\boldsymbol{\gamma}) \\ \overline{z}(t,\boldsymbol{p}) \end{bmatrix} \tag{5.17}$$

Let $\overline{\boldsymbol{H}} \in \mathbb{R}^{a,n-a}$ *given by*

$$\overline{\boldsymbol{H}}(t,\boldsymbol{p}) := \partial_{\boldsymbol{p}}\overline{\boldsymbol{h}}(t,\boldsymbol{p}) \tag{5.18}$$

have full rank, and let $\overline{\boldsymbol{P}} \in \mathbb{R}^{a,a}$ *defined by*

$$\overline{\boldsymbol{P}}(t,\boldsymbol{p}) := \overline{\boldsymbol{H}}(t,\boldsymbol{p})\boldsymbol{M}_1^{-1}(\boldsymbol{p})\boldsymbol{B}_1^T(\boldsymbol{p}) \tag{5.19}$$

be invertible.

Proposition 1. *Given Assumption 1, the DAEs* (5.14a)–(5.14e) *are of index 3.*

Proof. The idea of the proof is to reduce the system (5.14a)–(5.14e) to a system that has the structure of a constrained multibody system for which the index is known to be 3. At first the variables $\boldsymbol{x}$, $\widehat{\boldsymbol{x}}$, and $\widetilde{\boldsymbol{x}}$ can be eliminated since they are directly given by $\boldsymbol{\gamma}$ and its derivatives. Then the second part of equation (5.14a), namely

$$\boldsymbol{M}_2\ddot{\boldsymbol{\gamma}} = \boldsymbol{f}_2(\boldsymbol{\gamma},\dot{\boldsymbol{\gamma}}) - \boldsymbol{G}_2^T(\boldsymbol{p},\boldsymbol{\gamma})\boldsymbol{\lambda} \tag{5.20}$$

is used to extract an equation for $\boldsymbol{\lambda}$. Since $\boldsymbol{G}_2 \in \mathbb{R}^{m,a}$ is assumed to have full rank, Equation (5.20) yields

$$\boldsymbol{\lambda} = \left(\boldsymbol{G}_2(\boldsymbol{p},\boldsymbol{\gamma})\boldsymbol{G}_2^T(\boldsymbol{p},\boldsymbol{\gamma})\right)^{-1}\boldsymbol{G}_2(\boldsymbol{p},\boldsymbol{\gamma})\,(\boldsymbol{f}_2(\boldsymbol{\gamma},\dot{\boldsymbol{\gamma}}) - \boldsymbol{M}_2\ddot{\boldsymbol{\gamma}}) =: \overline{\boldsymbol{\lambda}}(t,\boldsymbol{p}) \tag{5.21}$$

In addition, premultiplying Equation (5.20) by $\boldsymbol{P}_2^T(\boldsymbol{p},\boldsymbol{\gamma})$ and taking into account Equation (5.15) give

$$\overline{z}(t,\boldsymbol{p}) = \boldsymbol{0} \tag{5.22}$$

where $\overline{z}(t,\boldsymbol{p})$ has been defined in Equation (5.16). Accordingly, the a equations in (5.20) yield m equations for the determination of $\boldsymbol{\lambda}(t,\boldsymbol{p})$ along with $a-m$ equations $\overline{z}(t,\boldsymbol{p}) = \boldsymbol{0}$, which can be viewed as additional algebraic constraints. To summarize, the system is eventually obtained, that is

$$M_1(p)\ddot{p} = \overline{f}_1(t,p,\dot{p}) + B_1^T(p)u \tag{5.23a}$$

$$0 = \overline{h}(t,p) \tag{5.23b}$$

where

$$\overline{f}_1(t,p,\dot{p}) := f_1(p,\dot{p}) - G_1^T(p,\gamma)\overline{\lambda}(t,p) \tag{5.24}$$

and $\overline{h}(t,p)$ has been defined in Equation (5.17). The DAEs (5.23a)–(5.23b) consist of $n-a$ differential equations (5.23a) and $m+(a-m)=a$ algebraic equations (5.23b) for the determination of $p \in \mathbb{R}^{n-a}$ and $u \in \mathbb{R}^a$. In particular, the DAEs (5.23a)–(5.23b) assume the semi-explicit structure known from multibody dynamics. More precisely, the DAEs (5.23a)–(5.23b) are Hessenberg index-3 (see, e.g., Ascher and Petzold [6, Sect. 9.1.1]). Provided that Assumption 1 holds, the DAEs (5.23a)–(5.23b) have index 3. To see this, it can be argued along the lines of Subsection 4.1.1. In particular, the argument hinges on the full rank assumption for the matrix $\overline{P}$ defined in Equation (5.19). □

Remark 2. *Proposition 1 implies that the original DAEs* (5.10a)–(5.10c) *have index 5 at most. This follows from the fact that two differentiation steps were sufficient to obtain DAEs of index 3.*

5.1.3 Reduction to index 1

The procedure of minimal extension can be applied a second time to eventually reach DAEs of index 1. However, due to the fact that the extended system (5.14a)–(5.14e) does not exhibit the desired Hessenberg form anymore, the index reduction method can not be directly applied to the DAEs (5.14a)–(5.14e). Then it is necessary to find the equations that need to be differentiated.

Here these equations have already been identified in the proof of Proposition 1. Accordingly, to apply index reduction by minimal extension a second time, the derivatives of the constraints (5.23b) need to be added. In this way, the original system is extended by $2a$ algebraic constraints. Correspondingly, $2a$ additional dummy variables need to be introduced to reach a square system. For this purpose, the first and second time derivative of the crane coordinates $p \in \mathbb{R}^{n-a}$ are available . That is, there are $2(n-a)$ variables at the disposal. Note that this complies with the relation $a \leqslant n-a$ in Assumption 1. Although the second index reduction can be performed for the general case $a \leqslant n-a$, the focus will be placed on the special case $a = n-a$ in the following part.

The special case of purely algebraic equations

In the sequel the special case $a = n - a$ is mainly considered and it applies to the numerical examples dealt with in Subsection 5.3.1 and 5.3.2. In this case the introduction of dummy derivatives implies that all differential variables in DAEs (5.23a)–(5.23b) are converted to algebraic ones. Thus, after the second index reduction, no differential variables are present any more, and the resulting system of equations is purely algebraic. This indicates that the specific systems under consideration are classified as differentially flat systems.

Provided that $a = n - a$, $\widehat{\boldsymbol{p}} := \dot{\boldsymbol{p}}$ and $\widetilde{\boldsymbol{p}} := \ddot{\boldsymbol{p}}$ are introduced as additional dummy variables. Eventually the system reads

$$\boldsymbol{M}_1(\boldsymbol{p})\widetilde{\boldsymbol{p}} = \overline{\boldsymbol{f}}_1(t,\boldsymbol{p},\widehat{\boldsymbol{p}}) + \boldsymbol{B}_1^T(\boldsymbol{p})\boldsymbol{u}, \tag{5.25a}$$

$$\boldsymbol{0} = \overline{\boldsymbol{h}}(t,\boldsymbol{p}) \tag{5.25b}$$

$$\boldsymbol{0} = \overline{\boldsymbol{H}}(t,\boldsymbol{p})\widehat{\boldsymbol{p}} + \partial_t\overline{\boldsymbol{h}}(t,\boldsymbol{p}) \tag{5.25c}$$

$$\boldsymbol{0} = \overline{\boldsymbol{H}}(t,\boldsymbol{p})\widetilde{\boldsymbol{p}} + \boldsymbol{\eta}(t,\boldsymbol{p},\widehat{\boldsymbol{p}}) \tag{5.25d}$$

where the i-th component of the vector-valued function $\boldsymbol{\eta}(t,\boldsymbol{p},\widehat{\boldsymbol{p}})$ is given by

$$\eta_i(t,\boldsymbol{p},\widehat{\boldsymbol{p}}) = \widehat{\boldsymbol{p}}^T\partial^2_{\boldsymbol{pp}}\overline{h}_i(t,\boldsymbol{p})\widehat{\boldsymbol{p}} + 2\partial^2_{t\boldsymbol{p}}\overline{h}_i(t,\boldsymbol{p})\widehat{\boldsymbol{p}} + \partial^2_{tt}\overline{h}_i(t,\boldsymbol{p}) \tag{5.26}$$

for $i = 1,\ldots,a$. Since system (5.25a)–(5.25d) is purely algebraic, it is easy to see that the DAEs have index 1. In particular, system (5.25a)–(5.25d) constitutes $4a$ algebraic equations for the determination of the $4a$ variables $\boldsymbol{p}$, $\widehat{\boldsymbol{p}}$, $\widetilde{\boldsymbol{p}}$ and $\boldsymbol{u}$.

Remark 3. *The assumptions made in Assumption 1 guarantee the unique solvability of the algebraic system* (5.25a)–(5.25d).

Remark 4. *Alternatively, the above extension procedure can also be applied directly to system* (5.14a)–(5.14e). *In this case the new constraints* (5.25c) *and* (5.25d) *need to be appended to the DAEs* (5.14a)– (5.14e). *In addition, the dummy variabels* $\widehat{\boldsymbol{p}} := \dot{\boldsymbol{p}}$ *and* $\widetilde{\boldsymbol{p}} := \ddot{\boldsymbol{p}}$ *need to be introduced. Again a purely algebraic system of equations, which is equivalent to system* (5.25a)–(5.25d), *is obtained.*

Remark 5. *A careful inspection of the present index-1 formulation shows that all unknowns (redundant coordinates, Lagrange multipliers, and control inputs) can be expressed in terms of the flat output function $\gamma(t)$ along with the derivatives thereof up to the fourth order.*

This corresponds to the fact that the crane models under consideration can be classified as differentially flat systems (see [26, 33, 82]).

Remark 6. *As mentioned before, the second index reduction can also be performed for the case* $n - a > a$*. In this case the introduction of dummy derivatives still leaves differential variables in the resulting index-1 DAEs. This is indicative for systems with internal dynamics (or zero dynamics). In this case additional issues may arise such as the stability of the internal dynamics.*

5.1.4 Minimal coordinates

The minimal extension procedure of index reduction can also be applied to crane formulations in terms of minimal coordinates. Based on the minimally extended index-3 formulation (5.14a)–(5.14e), the redundant coordinates (5.11) are expressed in terms of minimal coordinates $\boldsymbol{\mu} \in \mathbb{R}^f$ with $f = n - m$. Thus

$$q = \boldsymbol{\varphi}(\boldsymbol{\mu}) \qquad \text{or} \qquad \begin{bmatrix} p \\ x \end{bmatrix} = \begin{bmatrix} \varphi_1(\boldsymbol{\mu}) \\ \varphi_2(\boldsymbol{\mu}) \end{bmatrix} \tag{5.27}$$

Note that by definition the coordinate mapping[3] (5.27) satisfies identically the holonomic constraints (5.14b), that is, $\boldsymbol{\Phi} \circ \boldsymbol{\varphi}(\boldsymbol{\mu}) = \mathbf{0}$ for all $\boldsymbol{\mu} \subset \mathbb{R}^f$. In Subsection 5.3.1 and 5.3.2, the coordinate mapping (5.27) will be described in detail in the context of the specific examples.

Using the redundant coordinates (5.11), it is natural to select the derivatives of the load coordinates $\boldsymbol{x} \in \mathbb{R}^a$ as dummy variables (cf. Subsection 5.1.2). Similarly, differentiating the minimal coordinates twice with respect to time leads to the corresponding velocities $\dot{\boldsymbol{\mu}} \in \mathbb{R}^f$ and accelerations $\ddot{\boldsymbol{\mu}} \in \mathbb{R}^f$, from which appropriate dummy variables need to be selected. To this end, the minimal coordinates are split into $\boldsymbol{\mu}_1 \in \mathbb{R}^{f-a}$ and $\boldsymbol{\mu}_2 \in \mathbb{R}^a$ such that

$$D_2\varphi_2(\boldsymbol{\mu}_1, \boldsymbol{\mu}_2) \in \mathbb{R}^{a,a} \quad \text{is nonsingular.} \tag{5.28}$$

With a slight abuse of notation, $\boldsymbol{\varphi}(\boldsymbol{\mu}_1, \boldsymbol{\mu}_2)$ is used to express the mapping (5.27) after the coordinate partition has been performed. Furthermore, in Equation (5.28) and in the sequel, $D_\alpha\boldsymbol{\varphi}(\boldsymbol{\mu}_1, \boldsymbol{\mu}_2)$ with $\alpha = 1$ or $\alpha = 2$ denotes the partial derivative with respect to the first or second argument, respectively.

[3] Note that $\boldsymbol{\varphi}$ denotes here the mapping rather than the position vector used previously.

Now the dummy variables or dummy derivatives are chosen as

$$\widehat{\mu}_2 = \dot{\mu}_2 \tag{5.29a}$$
$$\widetilde{\mu}_2 = \ddot{\mu}_2 \tag{5.29b}$$

Differentiating the mapping (5.27) with respect to time leads to

$$\dot{p} = D_1\varphi_1(\mu_1, \mu_2)\dot{\mu}_1 + D_2\varphi_1(\mu_1, \mu_2)\widehat{\mu}_2 \tag{5.30a}$$
$$\ddot{p} = D_1\varphi_1(\mu_1, \mu_2)\ddot{\mu}_1 + D_2\varphi_1(\mu_1, \mu_2)\widetilde{\mu}_2 + g_1(\mu_1, \mu_2, \dot{\mu}_1, \widehat{\mu}_2) \tag{5.30b}$$

and

$$\widehat{x} = D_1\varphi_2(\mu_1, \mu_2)\dot{\mu}_1 + D_2\varphi_2(\mu_1, \mu_2)\widehat{\mu}_2 \tag{5.31a}$$
$$\widetilde{x} = D_1\varphi_2(\mu_1, \mu_2)\ddot{\mu}_1 + D_2\varphi_2(\mu_1, \mu_2)\widetilde{\mu}_2 + g_2(\mu_1, \mu_2, \dot{\mu}_1, \widehat{\mu}_2) \tag{5.31b}$$

where

$$g_\alpha(\mu_1, \mu_2, \dot{\mu}_1, \widehat{\mu}_2) = \frac{\mathrm{d}}{\mathrm{dt}}\left(D_1\varphi_\alpha(\mu_1, \mu_2)\right)\dot{\mu}_1 + \frac{\mathrm{d}}{\mathrm{dt}}\left(D_2\varphi_\alpha(\mu_1, \mu_2)\right)\widehat{\mu}_2 \tag{5.32}$$

These relationships can now be inserted into the minimally extended index-3 DAEs (5.14a)–(5.14e). In addition, in order to eliminate the Lagrange multipliers $\boldsymbol{\lambda}$ from (5.14a), Equation (5.14a) is multiplied from the left by $D\varphi(\mu)^T$. A straightforward calculation yields the minimally extended index-3 formulation in terms of minimal coordinates given by

$$\mathcal{M}_{11}(\mu)\ddot{\mu}_1 = h_1(\mu, \dot{\mu}_1, \widehat{\mu}_2) - \mathcal{M}_{12}(\mu)\widetilde{\mu}_2 - \mathcal{B}_1^T(\mu)u \tag{5.33a}$$
$$\mathcal{M}_{21}(\mu)\ddot{\mu}_1 = h_2(\mu, \dot{\mu}_1, \widehat{\mu}_2) - \mathcal{M}_{22}(\mu)\widetilde{\mu}_2 - \mathcal{B}_2^T(\mu)u \tag{5.33b}$$
$$D_1\varphi_2(\mu_1, \mu_2)\ddot{\mu}_1 = \ddot{\gamma} - g_2(\mu, \dot{\mu}_1, \widehat{\mu}_2) - D_2\varphi_2(\mu_1, \mu_2)\widetilde{\mu}_2 \tag{5.33c}$$
$$0 = D_1\varphi_2(\mu_1, \mu_2)\dot{\mu}_1 + D_2\varphi_2(\mu_1, \mu_2)\widehat{\mu}_2 - \dot{\gamma} \tag{5.33d}$$
$$0 = \varphi_2(\mu) - \gamma \tag{5.33e}$$

Here the components are given by

$$\mathcal{M}_{\alpha\beta}(\mu) = \sum_{\gamma=1}^{2} D_\alpha\varphi_\gamma^T(\mu_1, \mu_2)M_\gamma D_\beta\varphi_\gamma(\mu_1, \mu_2) \tag{5.34a}$$
$$h_\alpha(\mu, \dot{\mu}_1, \widehat{\mu}_2) = \sum_{\gamma=1}^{2} D_\alpha\varphi_\gamma^T(\mu_1, \mu_2)\left(f_\gamma - M_\gamma g_\gamma\right) \tag{5.34b}$$
$$\mathcal{B}_\alpha^T(\mu) = D_\alpha\varphi_1^T(\mu_1, \mu_2)B_1^T \tag{5.34c}$$

Note that, to simplify the notation, (μ_1, μ_2) has often been replaced by μ as an argument of the functions considered. Similarly, the arguments of functions pertaining to the underlying formulation in terms of redundant coordinates have been suppressed. The above procedure coincides with the reduced formulation of the DAEs (Subsection 2.3.2). Besides, the minimal coordinates formulation can also be obtained by applying Lagrange's equations of the second kind (see Section 2.2).

System (5.33a)−(5.33e) constitutes a set of $f + 3a$ index-3 DAEs for the determination of the differential variables $\mu_1 \in \mathbb{R}^{f-a}$ and the algebraic variables u, μ_2, $\widehat{\mu}_2$, $\widetilde{\mu}_2 \in \mathbb{R}^a$.

Commutative process

Minimal coordinates can also be employed from the outset, prior to the index reduction approach. Indeed, the index reduction by minimal extension may also start from the formulation in terms of minimal coordinates given by the system (4.1a)−(4.1c).

1. In a first step the coordinate mapping (5.27) is employed to convert the index-5 formulation in terms of redundant coordinates (5.10a)−(5.10c) to the corresponding index-5 formulation in terms of minimal coordinates (4.1a)−(4.1c). This conversion is a standard procedure relying on the projection matrix $D\varphi(\mu)$. Thus, the index-5 DAEs are obtained and read

$$\mathcal{M}(\mu)\ddot{\mu} = h(\mu, \dot{\mu}) - \mathcal{B}^T(\mu)u \tag{5.35a}$$

$$0 = \varphi_2(\mu) - \gamma \tag{5.35b}$$

where the reduced mass matrix $\mathcal{M}(\mu) = D\varphi^T(\mu)MD\varphi(\mu)$ assumes the partitioned form

$$\mathcal{M}(\mu) = \begin{bmatrix} \mathcal{M}_{11}(\mu) & \mathcal{M}_{12}(\mu) \\ \mathcal{M}_{21}(\mu) & \mathcal{M}_{22}(\mu) \end{bmatrix} \tag{5.36}$$

Here, the submatrices are given by Equation (5.34a). Similarly, h and $\mathcal{B}$ in Equation (5.35a) can be assembled from Equation (5.34b) and (5.34c), respectively. Note that comparing the servo constraints (5.35b) with (4.1c) shows that $\varphi_2(\mu) = s(\mu)$.

2. Now index reduction by minimal extension can be applied to the system of equations (5.35a)–(5.35b). To this end, partition the minimal coordinates subject to condition (5.28), differentiate the servo constraints (5.35b) twice with respect to time, and introduce the dummy variables (5.29a)–(5.29b). It is easy to see that this procedure yields again the index-3 DAEs (5.33a)–(5.33e).

Obviously, the two steps to arrive at the minimally extended set of index-3 DAEs in terms of minimal coordinates (5.33a)–(5.33e) do commute. That is, the final result is independent of the order of the steps (i) minimal extension and (ii) introduction of minimal coordinates. This is summarized in the commutative diagram in Fig. 5.1.

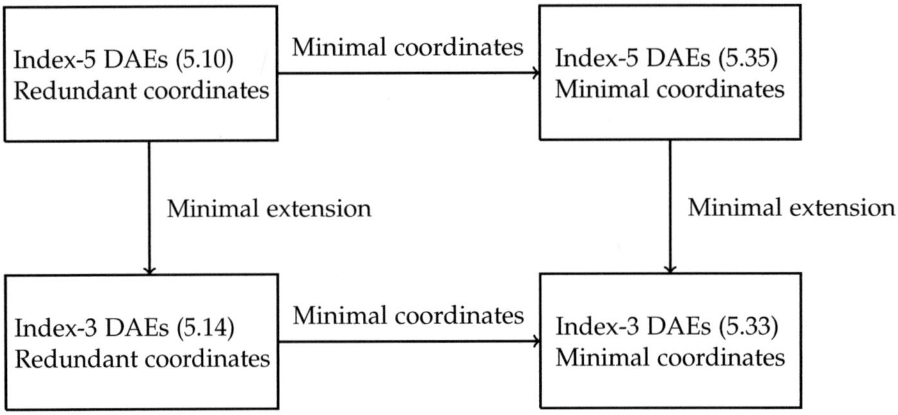

Figure 5.1: Commutative diagram for index reduction and the introduction of minimal coordinates.

Remark 7. *An alternative way of reducing the index from 5 to 3 is the projection method originally proposed by Blajer and Kołodziejczyk [27]. This approach requires the design of a suitable projection matrix and eventually yields $f + a$ index-3 DAEs. Whereas in the present approach the servo constraints are enforced on position, velocity and acceleration level (see* (5.33e), (5.33d) *and* (5.33c)*), the projection method enforces the servo constraints only on position and acceleration level. Correspondingly, the present approach is characterized by $f + 3a$ index-3 DAEs.*

5.2 Discretization

After the index of the equations of motion has been reduced to three, it is necessary to discuss the temporal discretization for the numerical simulation. For general DAEs of index 3, the stability of the used numerical integration method needs special attention. However, the semi-explicit form allows to apply the backward Euler scheme. Here the simple structure of the system, which is obtained by the minimal extension procedure, is beneficial to the following time discretization.

5.2.1 Index-3 formulation in terms of dependent coordinates

The minimally extended index-3 formulation in terms of redundant (dependent) coordinates (5.14a)–(5.14e) can be recast in the form

$$\boldsymbol{M}_1(\boldsymbol{p})\ddot{\boldsymbol{p}} = \boldsymbol{f}_1(\boldsymbol{p},\dot{\boldsymbol{p}}) + \boldsymbol{B}_1^T(\boldsymbol{p})\boldsymbol{u} - \boldsymbol{G}_1^T(\boldsymbol{p},\boldsymbol{\gamma})\boldsymbol{\lambda} \tag{5.37a}$$

$$\boldsymbol{0} = \boldsymbol{M}_2\ddot{\boldsymbol{\gamma}} - \boldsymbol{f}_2(\boldsymbol{\gamma},\dot{\boldsymbol{\gamma}}) + \boldsymbol{G}_2^T(\boldsymbol{p},\boldsymbol{\gamma})\boldsymbol{\lambda} \tag{5.37b}$$

$$\boldsymbol{0} = \boldsymbol{\Phi}(\boldsymbol{p},\boldsymbol{\gamma}) \tag{5.37c}$$

The DAEs (5.37a)–(5.37c) provide $n-a$ differential equations (5.37a) along with $a+m$ algebraic equations (5.37b) and (5.37c) for the determination of $\boldsymbol{p} \in \mathbb{R}^{n-a}$, $\boldsymbol{u} \in \mathbb{R}^a$, and $\boldsymbol{\lambda} \in \mathbb{R}^m$. In particular, the DAEs (5.37a)–(5.37c) are in semi-explicit form, so that the simple backward Euler discretization is expected to work well (see Ascher and Petzold [6, Sec. 10.1.1]). Accordingly, the scheme

$$\boldsymbol{p}_{n+1} - \boldsymbol{p}_n = \Delta t \boldsymbol{v}_{n+1} \tag{5.38a}$$

$$\begin{aligned}&\boldsymbol{M}_1(\boldsymbol{p}_{n+1})\left(\boldsymbol{v}_{n+1} - \boldsymbol{v}_n\right)\\ &\quad = \Delta t\left(\boldsymbol{f}_1(\boldsymbol{p}_{n+1},\boldsymbol{v}_{n+1}) + \boldsymbol{B}_1^T(\boldsymbol{p}_{n+1})\boldsymbol{u}_{n+1} - \boldsymbol{G}_1^T(\boldsymbol{p}_{n+1},\boldsymbol{\gamma}(t_{n+1}))\boldsymbol{\lambda}_{n+1}\right)\end{aligned} \tag{5.38b}$$

$$\boldsymbol{0} = \boldsymbol{M}_2\ddot{\boldsymbol{\gamma}}(t_{n+1}) - \boldsymbol{f}_2(\boldsymbol{\gamma}(t_{n+1}),\dot{\boldsymbol{\gamma}}(t_{n+1})) + \boldsymbol{G}_2^T(\boldsymbol{p}_{n+1},\boldsymbol{\gamma}(t_{n+1}))\boldsymbol{\lambda}_{n+1} \tag{5.38c}$$

$$\boldsymbol{0} = \boldsymbol{\Phi}(\boldsymbol{p}_{n+1}\boldsymbol{\gamma}(t_{n+1})) \tag{5.38d}$$

is considered. In a typical step of size $\Delta t = t_{n+1} - t_n$ approximations $(\bullet)_{n+1}$ to $(\bullet)(t_{n+1})$ need to be found if the corresponding quantities $(\bullet)_n$ are given as the result of the previous step. For the initial step, consistent initial values $\boldsymbol{p}_0$ and $\boldsymbol{v}_0$ are required and they need to satisfy $\boldsymbol{\Phi}(\boldsymbol{p}_0,\boldsymbol{\gamma}(t_0)) = \boldsymbol{0}$ along with

$$G_1(p_0, \gamma(t_0))v_0 + G_2(p_0, \gamma(t_0))\dot{\gamma}(t_0)) = \mathbf{0} \tag{5.39}$$

The scheme (5.38a)–(5.38d) provides $2n + m - a$ algebraic equations for the determination of $p_{n+1}, v_{n+1} \in \mathbb{R}^{n-a}$, $u_{n+1} \in \mathbb{R}^a$, and $\lambda_{n+1} \in \mathbb{R}^m$.

5.2.2 Index-3 formulation in terms of minimal coordinates

For the minimally extended index-3 formulation in terms of minimal coordinates (5.33a)– (5.33e), the backward Euler discretization can also be applied.

The corresponding scheme is given by

$$\mu_{1_{n+1}} - \mu_{1_n} = \Delta t \nu_{1_{n+1}} \tag{5.40a}$$

$$\begin{aligned}\mathcal{M}_{11}(\mu_{n+1})\left(\nu_{1_{n+1}} - \nu_{1_n}\right)\\ = \Delta t \left(h_1(\mu_{n+1}, \nu_{1_{n+1}}, \widehat{\mu}_{2_{n+1}}) - \mathcal{M}_{12}(\mu_{n+1})\widetilde{\mu}_{2_{n+1}} - \mathcal{B}_1^T(\mu_{n+1})u_{n+1}\right)\end{aligned} \tag{5.40b}$$

$$\begin{aligned}\mathcal{M}_{21}(\mu_{n+1})\left(\nu_{1_{n+1}} - \nu_{1_n}\right)\\ = \Delta t \left(h_2(\mu_{n+1}, \nu_{1_{n+1}}, \widehat{\mu}_{2_{n+1}}) - \mathcal{M}_{22}(\mu_{n+1})\widetilde{\mu}_{2_{n+1}} - \mathcal{B}_2^T(\mu_{n+1})u_{n+1}\right)\end{aligned} \tag{5.40c}$$

$$\begin{aligned}D_1\varphi_2(\mu_{1_{n+1}}, \mu_{2_{n+1}})\left(\nu_{1_{n+1}} - \nu_{1_n}\right)\\ = \Delta t \left(\ddot{\gamma}(t_{n+1}) - g_2(\mu_{n+1}, \nu_{1_{n+1}}, \widehat{\mu}_{2_{n+1}}) - D_2\varphi_2(\mu_{1_{n+1}}, \mu_{2_{n+1}})\widetilde{\mu}_{2_{n+1}}\right)\end{aligned} \tag{5.40d}$$

$$\mathbf{0} = D_1\varphi_2(\mu_{1_{n+1}}, \mu_{2_{n+1}})\nu_{1_{n+1}} + D_2\varphi_2(\mu_{1_{n+1}}, \mu_{2_{n+1}})\widehat{\mu}_{2_{n+1}} - \dot{\gamma}(t_{n+1}) \tag{5.40e}$$

$$\mathbf{0} = \varphi_2(\mu_{n+1}) - \gamma(t_{n+1}) \tag{5.40f}$$

The scheme (5.40a)–(5.40f) provides $2(f + a)$ algebraic equations for the determination of $\mu_{1_{n+1}}, \nu_{1_{n+1}} \in \mathbb{R}^{f-a}$ and $\mu_{2_{n+1}}, \widehat{\mu}_{2_{n+1}}, \widetilde{\mu}_{2_{n+1}}, u_{n+1} \in \mathbb{R}^a$.

5.3 Numerical examples

Here the crane examples treated in Section 4.4 will be used again to demonstrate the application of index reduction by minimal extension to the crane formulations in terms of both redundant (dependent) and minimal coordinates.

5.3.1 Planar overhead crane

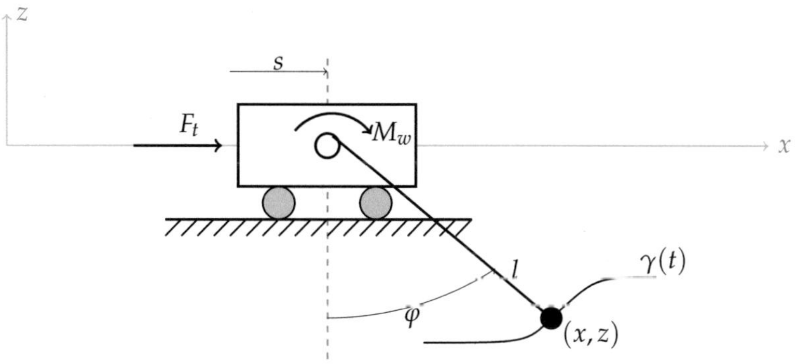

Figure 5.2: The model of the planar overhead crane.

As the first example, the planar overhead crane, that allows traveling and hoisting motions (see Fig. 5.2), is considered. This servo constraint problem has originally been formulated in terms of minimal coordinates in [27] and recast in redundant coordinates in [20, 30, 94] (see also Subsection 4.4.1).

The description of the overhead crane is based on $n = 4$ redundant coordinates, $m = 1$ holonomic constraint, and $a = 2$ controls. In particular, the crane coordinates $\boldsymbol{p} \in \mathbb{R}^2$ and the load coordinates $\boldsymbol{x} \in \mathbb{R}^2$ are given by

$$\boldsymbol{p} = \begin{bmatrix} s \\ l \end{bmatrix} \quad \text{and} \quad \boldsymbol{x} = \begin{bmatrix} x \\ z \end{bmatrix} \tag{5.41}$$

As before, the horizontal position of the trolley is given by s, the cable length is l, and the coordinates of the load are denoted by (x, z). The redundant coordinates need to satisfy the holonomic constraint

$$\Phi(\boldsymbol{p}, \boldsymbol{x}) = \frac{1}{2}\left((x - s)^2 + z^2 - l^2\right) = 0 \tag{5.42}$$

The holonomic constraint gives rise to the associated constraint Jacobian, which can be decomposed into

$$\boldsymbol{G}_1(\boldsymbol{p}, \boldsymbol{x}) = \begin{bmatrix} -(x - s) & -l \end{bmatrix} \tag{5.43a}$$

$$\boldsymbol{G}_2(\boldsymbol{p}, \boldsymbol{x}) = \begin{bmatrix} x - s & z \end{bmatrix} \tag{5.43b}$$

In addition, the underlying index-5 DAEs (5.10a)–(5.10c) employ the mass matrices

$$M_1 = \begin{bmatrix} m_t & 0 \\ 0 & \frac{J}{r^2} \end{bmatrix} \quad \text{and} \quad M_2 = \begin{bmatrix} m & 0 \\ 0 & m \end{bmatrix} \tag{5.44}$$

where the mass of the trolley is m_t, the moment of inertia of the winch is J, the winch radius is r, and the mass of the load is m. Further, the quantities needed in (5.10a)–(5.10c) are given by

$$f_1 = \begin{bmatrix} 0 \\ 0 \end{bmatrix}, \qquad f_2 = \begin{bmatrix} 0 \\ -mg \end{bmatrix}, \qquad B_1 = \begin{bmatrix} 1 & 0 \\ 0 & \frac{1}{r} \end{bmatrix} \tag{5.45}$$

The servo constraints (5.10c) are used to prescribe the trajectory of the load. Accordingly, there is

$$\gamma = \begin{bmatrix} x_d \\ z_d \end{bmatrix} \tag{5.46}$$

where the coordinates x_d and z_d are prescribed functions of time. The corresponding control inputs assume the form

$$u = \begin{bmatrix} F_t \\ M_w \end{bmatrix} \tag{5.47}$$

where the force acting on the trolley is given by F_t, and the torque acting on the winch is given by M_w.

Verification of Assumption 1

To verify Assumption 1, at first a matrix

$$P_2(p, \gamma) = \begin{bmatrix} -z_d \\ x_d - s \end{bmatrix} \tag{5.48}$$

is chosen such that condition (5.15) is satisfied. Then Equation (5.16) yields

$$\overline{z}(t, p) = m\left(z_d \ddot{x}_d - (x_d - s)(g + \ddot{z}_d)\right) \tag{5.49}$$

Furthermore, the constraint function (5.17) reads

$$\overline{h}(t, p) = \begin{bmatrix} \frac{1}{2}\left((x_d - s)^2 + z_d^2 - l^2\right) \\ m\left(z_d \ddot{x}_d - (x_d - s)(g + \ddot{z}_d)\right) \end{bmatrix} \tag{5.50}$$

such that Equation (5.18) yields

$$\overline{\boldsymbol{H}}(t,\boldsymbol{p}) = \begin{bmatrix} \boldsymbol{G}_1(\boldsymbol{p},\boldsymbol{\gamma}) \\ \partial_{\boldsymbol{p}}\bar{z}(t,\boldsymbol{p}) \end{bmatrix} = \begin{bmatrix} -(x_d - s) & -l \\ m(g+\ddot{z}_d) & 0 \end{bmatrix} \tag{5.51}$$

Eventually, Equation (5.19) gives

$$\boldsymbol{P}(t,\boldsymbol{p}) = \begin{bmatrix} -\frac{x_d - s}{m_t} & -\frac{lr}{J} \\ \frac{m}{m_t}(g+\ddot{z}_d) & 0 \end{bmatrix} \tag{5.52}$$

Note that in practical applications there are $l > 0$ and $g + \ddot{z}_d > 0$. The last inequality holds due to the fact that the cable (which in the present model is assumed to be inextensible and massless) connecting the load with the winch can only sustain tensile (and no compressive) forces. This can be easily verified by applying Newton's second law of motion. Thus, $\overline{\boldsymbol{H}}(t,\boldsymbol{p})$ has full rank, and $\overline{\boldsymbol{P}}(t,\boldsymbol{p})$ is invertible. Consequently, Assumption 1 is satisfied, and Proposition 1 holds.

It should be further noted that the minimally extended index-3 DAEs (5.23a)–(5.23b) can now be set up for the overhead crane. It only remains to calculate

$$\overline{\boldsymbol{f}}_1(t,\boldsymbol{p}) = \frac{m}{l^2}\left((x_d - s)\ddot{x}_d + z_d(g+\ddot{z}_d)\right)\begin{bmatrix} -(x_d - s) \\ -l \end{bmatrix} \tag{5.53}$$

to complete the description of the DAEs (5.23a)–(5.23b).

Index-1 formulation

As explained in Section 5.1.3, the index-1 formulation (5.25a)–(5.25d) yields a purely algebraic system of equations that facilitates an analytical solution to the inverse dynamics problem under consideration. The additional quantities needed in (5.25c) and (5.25d) read

$$\partial_t\overline{\boldsymbol{h}}(t,\boldsymbol{p}) = \begin{bmatrix} (x_d - s)\dot{x}_d + z_d\dot{z}_d \\ m\left(\dot{z}_d\ddot{x}_d - \dot{x}_d(g+\ddot{z}_d) + z_d x_d^{(3)} - (x_d - s)z_d^{(3)}\right) \end{bmatrix} \tag{5.54}$$

and also

$$\partial^2_{pp}\bar{h}_1(t,\boldsymbol{p}) = \begin{bmatrix} 1 & 0 \\ 0 & -1 \end{bmatrix} \tag{5.55a}$$

$$\partial^2_{pp}\bar{h}_2(t,\boldsymbol{p}) = \begin{bmatrix} 0 & 0 \\ 0 & 0 \end{bmatrix} \tag{5.55b}$$

$$\partial^2_{tp}\bar{\boldsymbol{h}}(t,\boldsymbol{p}) = \begin{bmatrix} -\dot{x}_d & 0 \\ mz_d^{(3)} & 0 \end{bmatrix} \tag{5.55c}$$

$$\partial^2_{tt}\bar{\boldsymbol{h}}(t,\boldsymbol{p}) = \begin{bmatrix} \dot{x}_d^2 + \dot{z}_d^2 + (x_d - s)\ddot{x}_d + z_d\ddot{z}_d \\ m\left(-\ddot{x}_d g + 2\dot{z}_d x_d^{(3)} - 2\dot{x}_d z_d^{(3)} + z_d x_d^{(4)} - (x_d - s) z_d^{(4)}\right) \end{bmatrix} \tag{5.55d}$$

In the present case it is possible to get a closed-form analytical solution to system (5.25a)– (5.25d), which serves as reference solution in the numerical simulation presented later. Note that the fourth order derivative of the prescribed output appears in the above index-1 formulation, and that means $\alpha = 4$. Thus, it proves that the original DAEs (5.10a)–(5.10c) have the index of five. It is identical to the flatness-based solution given in Subsection 4.4.1.

Minimal coordinates

Next the minimally extended index-3 system (5.33a)–(5.33e) will be considered for the overhead crane in terms of minimal coordinates. Since the planar overhead crane has $f = n - m = 3$ degrees of freedom, minimal coordinates are expressed by

$$\boldsymbol{\mu} = \begin{bmatrix} s \\ l \\ \varphi \end{bmatrix} \tag{5.56}$$

These coordinates have been also used in the original description of the present servo constraint problem in [27]. The coordinate mappings in (5.27) assume the form

$$\boldsymbol{\varphi}_1(\boldsymbol{\mu}) = \begin{bmatrix} s \\ l \end{bmatrix} \tag{5.57}$$

and also

$$\boldsymbol{\varphi}_2(\boldsymbol{\mu}) = \begin{bmatrix} s + l\sin\varphi \\ -l\cos\varphi \end{bmatrix} \tag{5.58}$$

For the minimal extension procedure, the minimal coordinates are split into

$$\mu_1 = \begin{bmatrix} s \end{bmatrix} \tag{5.59}$$

and

$$\mu_2 = \begin{bmatrix} l \\ \varphi \end{bmatrix} \tag{5.60}$$

such that the Jacobian

$$D_2\varphi_2(\mu_1, \mu_2) = \begin{bmatrix} \sin\varphi & l\cos\varphi \\ -\cos\varphi & l\sin\varphi \end{bmatrix} \tag{5.61}$$

is guaranteed to be nonsingular. Thus, condition (5.28) is satisfied. Furthermore,

$$D_1\varphi_1(\mu_1, \mu_2) = \begin{bmatrix} 1 \\ 0 \end{bmatrix}, \quad D_2\varphi_1(\mu_1, \mu_2) = \begin{bmatrix} 0 & 0 \\ 1 & 0 \end{bmatrix}, \quad D_1\varphi_2(\mu_1, \mu_2) = \begin{bmatrix} 1 \\ 0 \end{bmatrix} \tag{5.62}$$

Now, Equation (5.32) gives rise to

$$g_1 = \begin{bmatrix} 0 \\ 0 \end{bmatrix} \quad \text{and} \quad g_2 = \begin{bmatrix} 2\hat{\varphi}\hat{l}\cos\varphi - l\hat{\varphi}^2\sin\varphi \\ 2\hat{\varphi}\hat{l}\sin\varphi + l\hat{\varphi}^2\cos\varphi \end{bmatrix} \tag{5.63}$$

Note that the minimal extension procedure implies the equalities $\hat{l} = \dot{l}$ and $\hat{\varphi} = \dot{\varphi}$ due to the introduction of dummy variables $\hat{l}$ and $\hat{\varphi}$. Furthermore, (5.34a)–(5.34c) yields

$$\mathcal{M}_{11}(\mu) = \begin{bmatrix} m_t + m \end{bmatrix} \qquad \mathcal{M}_{12}(\mu) = \begin{bmatrix} m\sin\varphi & ml\cos\varphi \end{bmatrix} \tag{5.64a}$$

$$\mathcal{M}_{21}(\mu) = \mathcal{M}_{12}^T(\mu) \qquad \mathcal{M}_{22}(\mu) = \begin{bmatrix} \frac{J}{r^2} + m & 0 \\ 0 & ml^2 \end{bmatrix} \tag{5.64b}$$

and also

$$h_1(\mu, \dot{\mu}_1, \widehat{\mu}_2) = \begin{bmatrix} ml\hat{\varphi}^2\sin\varphi - 2m\hat{\varphi}\hat{l}\cos\varphi \end{bmatrix} \tag{5.65a}$$

$$h_2(\mu, \dot{\mu}_1, \widehat{\mu}_2) = \begin{bmatrix} ml\hat{\varphi}^2 + mg\cos\varphi \\ -ml(g\sin\varphi + 2\hat{\varphi}\hat{l}) \end{bmatrix} \tag{5.65b}$$

$$\mathcal{B}_1^T(\mu) = \begin{bmatrix} 1 & 0 \end{bmatrix} \tag{5.65c}$$

$$\mathcal{B}_2^T(\mu) = \begin{bmatrix} 0 & \frac{1}{r} \\ 0 & 0 \end{bmatrix} \tag{5.65d}$$

This completes the index-3 DAEs (5.33a)–(5.33e) for the overhead crane in terms of minimal coordinates. It is obviously seen that the above quantities are identical to the quantities that have been deduced in the projection method for the minimal coordinates formulation (see Subsection 4.4.1). Instead of differential variables, dummy variables appear in the minimal coordinates formulation here.

Inverse dynamics simulation

The data for the present numerical example have been taken from [27]. Accordingly, the prescribed trajectory of load m is defined by

$$\gamma(t) = \gamma_0 + (\gamma_f - \gamma_0)c(\tau) \tag{5.66}$$

with the initial position

$$\gamma_0 = \begin{bmatrix} x_d(t_0) \\ z_d(t_0) \end{bmatrix} = \begin{bmatrix} 0 \\ -4\,\mathrm{m} \end{bmatrix} \quad \text{at} \quad t_0 = 0 \tag{5.67}$$

and the final position

$$\gamma_f = \begin{bmatrix} x_d(t_f) \\ z_d(t_f) \end{bmatrix} = \begin{bmatrix} 5\,\mathrm{m} \\ -1\,\mathrm{m} \end{bmatrix} \quad \text{at} \quad t_f = 3\,\mathrm{s} \tag{5.68}$$

The interpolating polynomial $c(\tau)$ takes the form

$$c(\tau) = 70\tau^9 - 315\tau^8 + 540\tau^7 - 420\tau^6 + 126\tau^5 \quad \text{with} \quad \tau = \frac{t}{t_f - t_0} \tag{5.69}$$

Accordingly, the motion of load m is subjected to a rest-to-rest maneuver on a straight-line trajectory. Starting at rest, the initial configuration of the system is given by

$$q_0 = \begin{bmatrix} s_0 & l_0 & x_0 & z_0 \end{bmatrix}^T = \begin{bmatrix} 0 & 4\,\mathrm{m} & 0 & -4\,\mathrm{m} \end{bmatrix}^T \tag{5.70}$$

The remaining parameters are given by $m_t = 10\,\mathrm{kg}$, $m = 100\,\mathrm{kg}$, $J = 0.1\,\mathrm{kg}\cdot\mathrm{m}^2$, $g = 9.81\,\mathrm{m/s}^2$ and $r = 0.1\,\mathrm{m}$.

The simulation results for different time step sizes are depicted in Fig. 5.3 and 5.4. In each diagram, the numerical solution (NUM) is compared to the analytical reference solution (REF). It can be observed that the numerical solution converges to the analytical reference solution when the time step size is reduced. Here, the use of coordinates between redundant and minimal coordinates is not distinguished since both formulations yield very similar results. This also applies for the implementation of the projection method proposed in [27]. The motion of the overhead crane is illustrated in Fig. 5.5 with some snapshots at consecutive points in time. It shows that the snapshots are the same as in Subsection 4.4.1.

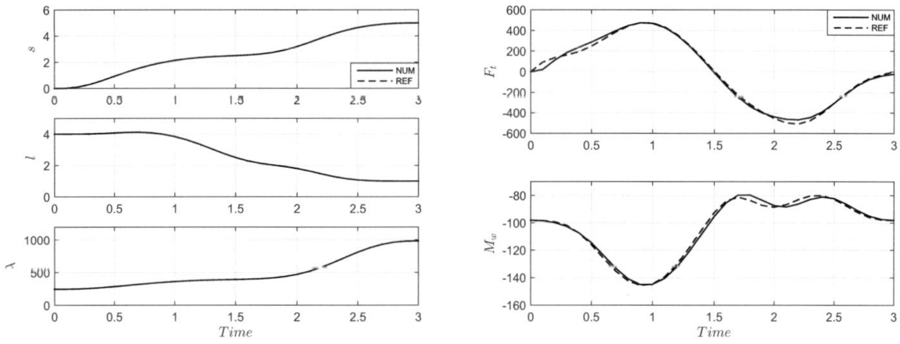

Figure 5.3: Planar overhead crane: Comparison between the numerical results (NUM) obtained with $\Delta t = 0.1$ s and the analytical reference solution (REF).

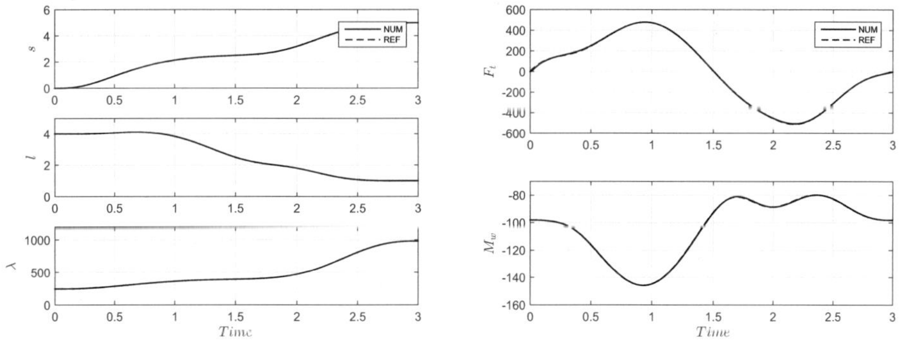

Figure 5.4: Planar overhead crane: Comparison between the numerical results (NUM) obtained with $\Delta t = 0.01$ s and the analytical reference solution (REF).

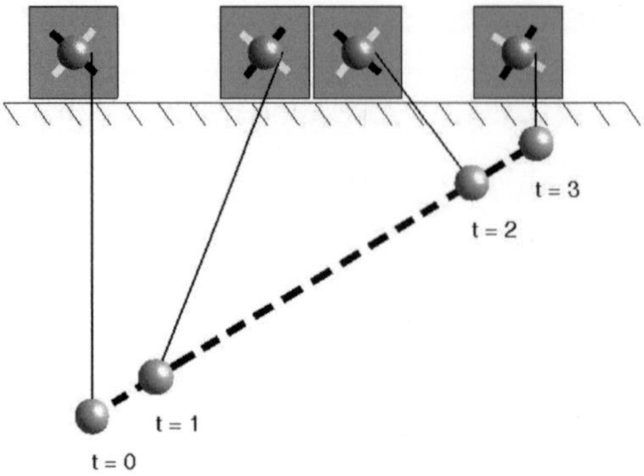

Figure 5.5: Planar overhead crane: Snapshots at specific points in time.

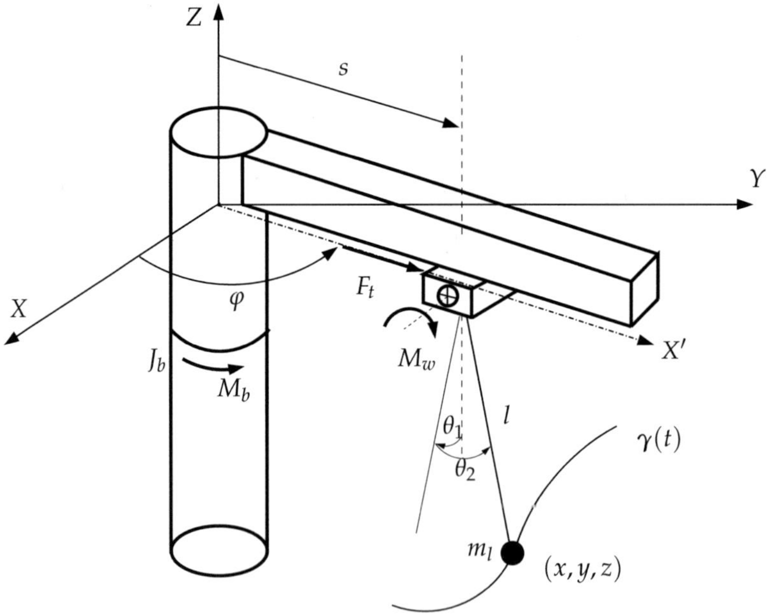

Figure 5.6: The model of the three-dimensional rotary crane.

5.3.2 Three-dimensional rotary crane

As the second example depicted in Fig. 5.6, the three-dimensional rotary crane treated in the forward dynamics simulation in Section 2.4, will be treated in the context of servo constraint problems again. This servo constraint problem has originally been dealt with in [31].

This problem can also be viewed as a 3d extension of the planar crane treated in the previous subsection. The 3d crane makes use of $n = 6$ redundant (dependent) coordinates that are subject to $m = 1$ holonomic constraint. Moreover, $a = 3$ servo constraints are used to prescribe the trajectory of the load. The crane coordinates $\boldsymbol{p} \in \mathbb{R}^3$ and the load coordinates $\boldsymbol{x} \in \mathbb{R}^3$ are given by

$$\boldsymbol{p} = \begin{bmatrix} \varphi & s & l \end{bmatrix}^T \quad \text{and} \quad \boldsymbol{x} = \begin{bmatrix} x & y & z \end{bmatrix}^T \tag{5.71}$$

The position of the load (mass m) is specified by the Cartesian coordinates (x, y, z) relative to an inertial reference frame. In addition to the location s of the trolley and the length l of the hoisting cable, the angle φ measures the rotation of the girder bridge about the Z-axis relative to the X-axis. Accordingly, the motion of the suspension point is described by polar coordiantes (s, φ) relative to the origin of the reference frame. The redundant coordinates need to satisfy the holonomic constraint

$$\Phi(\boldsymbol{p}, \boldsymbol{x}) = \frac{1}{2}\left((x - s\cos\varphi)^2 + (y - s\sin\varphi)^2 + z^2 - l^2\right) = 0 \tag{5.72}$$

The associated constraint Jacobian assumes the partitioned form

$$\boldsymbol{G}_1(\boldsymbol{p}, \boldsymbol{x}) = \begin{bmatrix} (x\sin\varphi - y\cos\varphi)s & (s - x\cos\varphi - y\sin\varphi) & -l \end{bmatrix} \tag{5.73a}$$

$$\boldsymbol{G}_2(\boldsymbol{p}, \boldsymbol{x}) = \begin{bmatrix} (x - s\cos\varphi) & (y - s\sin\varphi) & z \end{bmatrix} \tag{5.73b}$$

In the underlying index-5 DAEs (5.10a)–(5.10c), the mass matrices are given by

$$\boldsymbol{M}_1 = \begin{bmatrix} J_b + m_t s^2 & 0 & 0 \\ 0 & m_t & 0 \\ 0 & 0 & \frac{J_w}{r_w^2} \end{bmatrix}, \qquad \boldsymbol{M}_2 = \begin{bmatrix} m & 0 & 0 \\ 0 & m & 0 \\ 0 & 0 & m \end{bmatrix} \tag{5.74}$$

where the inertia value J_b is the moment of inertia of the bridge relative to the Z-axis, the inertia value J_w is the moment of inertia of the winch (of radius r_w), and the mass of the trolley is m_t. Further, the quantities needed in (5.10a)–(5.10c) are given by

$$f_1 = \begin{bmatrix} 2m_t s\dot{s}\dot{\varphi} \\ -m_t s\dot{\varphi}^2 \\ 0 \end{bmatrix} \qquad f_2 = \begin{bmatrix} 0 \\ 0 \\ -mg \end{bmatrix} \qquad \boldsymbol{B}_1 = \begin{bmatrix} 1 & 0 & 0 \\ 0 & 1 & 0 \\ 0 & 0 & \frac{1}{r_w} \end{bmatrix} \tag{5.75}$$

The servo constraints (5.10c) are used to prescribe the trajectory of the load. Accordingly, there is

$$\gamma = \begin{bmatrix} x_d & y_d & z_d \end{bmatrix}^T \tag{5.76}$$

where the coordinates x_d, y_d and z_d are prescribed functions of time. The control inputs assume the form

$$\boldsymbol{u} = \begin{bmatrix} M_b & F_t & M_w \end{bmatrix}^T \tag{5.77}$$

Here, the variable M_b is the torque acting about the $Z-$axis on the bridge, the variable F_t is the force acting along the girder bridge on the trolley, and the variable M_w is the torque acting on the winch.

Verification of Assumption 1

To verify Assumption 1, at first a matrix

$$\boldsymbol{P}_2(\boldsymbol{p}, \gamma) = \begin{bmatrix} -z_d & 0 \\ 0 & -z_d \\ x_d - s\cos\varphi & y_d - s\sin\varphi \end{bmatrix} \tag{5.78}$$

is chosen such that condition (5.15) is satisfied. Now Equation (5.16) yields

$$\bar{z}(t, \boldsymbol{p}) = \begin{bmatrix} m\,(z_d\ddot{x}_d - (x_d - s\cos\varphi)(g + \ddot{z}_d)) \\ m\,(z_d\ddot{y}_d - (y_d - s\sin\varphi)(g + \ddot{z}_d)) \end{bmatrix} \tag{5.79}$$

Furthermore, the constraint function (5.17) reads

$$\bar{\boldsymbol{h}}(t, \boldsymbol{p}) = \begin{bmatrix} \frac{1}{2}\left((x_d - s\cos\varphi)^2 + (y_d - s\sin\varphi)^2 + z_d^2 - l^2\right) \\ m\,(z_d\ddot{x}_d - (x_d - s\cos\varphi)(g + \ddot{z}_d)) \\ m\,(z_d\ddot{y}_d - (y_d - s\sin\varphi)(g + \ddot{z}_d)) \end{bmatrix} \tag{5.80}$$

such that Equation (5.18) yields

$$\overline{\boldsymbol{H}}(t,\boldsymbol{p}) = \begin{bmatrix} \boldsymbol{G}_1(\boldsymbol{p},\boldsymbol{\gamma}) \\ \partial_{\boldsymbol{p}}\overline{\boldsymbol{z}}(t,\boldsymbol{p}) \end{bmatrix} = \begin{bmatrix} (x_d \sin\varphi - y_d \cos\varphi)s & (s - x_d \cos\varphi - y_d \sin\varphi) & -l \\ -ms(g+\ddot{z}_d)\sin\varphi & m(g+\ddot{z}_d)\cos\varphi & 0 \\ ms(g+\ddot{z}_d)\cos\varphi & m(g+\ddot{z}_d)\sin\varphi & 0 \end{bmatrix} \tag{5.81}$$

Eventually, Equation (5.19) gives

$$\overline{\boldsymbol{P}}(t,\boldsymbol{p}) = \begin{bmatrix} \frac{(x_d \sin\varphi - y_d\cos\varphi)s}{J_b + m_t s^2} & \frac{s - x_d\cos\varphi - y_d\sin\varphi}{m_t} & -\frac{l r_w}{J_w} \\ -\frac{ms(g+\ddot{z}_d)\sin\varphi}{J_b+m_t s^2} & \frac{m(g+\ddot{z}_d)\cos\varphi}{m_t} & 0 \\ \frac{ms(g+\ddot{z}_d)\cos\varphi}{J_b+m_t s^2} & \frac{m(g+\ddot{z}_d)\sin\varphi}{m_t} & 0 \end{bmatrix} \tag{5.82}$$

As in the case of the planar overhead crane, the hoisting cable can only sustain tensile forces such that $g + \ddot{z}_d > 0$. Moreover, in practical applications, $l > 0$. This implies that $\overline{\boldsymbol{H}}(t,\boldsymbol{p})$ has full rank and $\overline{\boldsymbol{P}}(t,\boldsymbol{p})$ is invertible. Consequently, Assumption 1 is satisfied, and Proposition 1 holds.

It should be further noted that the minimally extended index-3 DAEs (5.23a)–(5.23b) can now be set up for the rotary crane. To complete the description of the DAEs (5.23a)–(5.23b), it only remains to calculate

$$\begin{aligned} \overline{\boldsymbol{f}}_1(t,\boldsymbol{p}) = & \begin{bmatrix} 2m_t s\dot{s}\dot{\varphi} \\ -m_t s\dot{\varphi}^2 \\ 0 \end{bmatrix} + \frac{m}{l^2}\left((x_d - s\cos\varphi)\ddot{x}_d + (y_d - s\sin\varphi)\ddot{y}_d + z_d(g+\ddot{z}_d)\right) \\ & \times \begin{bmatrix} (x_d \sin\varphi - y_d\cos\varphi)s \\ s - x_d\cos\varphi - y_d\sin\varphi \\ -l \end{bmatrix} \end{aligned} \tag{5.83}$$

Index-1 formulation

As explained in Section 5.1.3, the index-1 formulation (5.25a)–(5.25d) of the three-dimensional rotary crane model yields a purely algebraic system of equations, which facilitates to provide an analytical solution to the inverse dynamics simulation problem under consideration.

The additional quantities needed in (5.25c) and (5.25d) read

$$\partial_t \bar{\boldsymbol{h}}(t,\boldsymbol{p}) = \begin{bmatrix} (x_d - s\cos\varphi)\dot{x}_d + (y_d - s\sin\varphi)\dot{y}_d + z_d\dot{z}_d \\ m\left(z_d x_d^{(3)} + \dot{z}_d\ddot{x}_d - (x_d - s\cos\varphi)z_d^{(3)} - \dot{x}_d(g+\ddot{z}_d)\right) \\ m\left(\dot{z}_d\ddot{y}_d + z_d y_d^{(3)} - (y_d - s\sin\varphi)z_d^{(3)} - \dot{y}_d(g+\ddot{z}_d)\right) \end{bmatrix} \tag{5.84}$$

and

$$\partial^2_{pp}\bar{h}_1(t,\boldsymbol{p}) = \begin{bmatrix} (x_d\cos\varphi + y_d\sin\varphi)s & x_d\sin\varphi - y_d\cos\varphi & 0 \\ x_d\sin\varphi - y_d\cos\varphi & 1 & 0 \\ 0 & 0 & -1 \end{bmatrix} \tag{5.85a}$$

$$\partial^2_{pp}\bar{h}_2(t,\boldsymbol{p}) = \begin{bmatrix} -ms(g+\ddot{z}_d)\cos\varphi & -m(g+\ddot{z}_d)\sin\varphi & 0 \\ -m(g+\ddot{z}_d)\sin\varphi & 0 & 0 \\ 0 & 0 & 0 \end{bmatrix} \tag{5.85b}$$

$$\partial^2_{pp}\bar{h}_3(t,\boldsymbol{p}) = \begin{bmatrix} -ms(g+\ddot{z}_d)\sin\varphi & m(g+\ddot{z}_d)\cos\varphi & 0 \\ m(g+\ddot{z}_d)\cos\varphi & 0 & 0 \\ 0 & 0 & 0 \end{bmatrix} \tag{5.85c}$$

$$\partial^2_{tp}\bar{\boldsymbol{h}}(t,\boldsymbol{p}) = \begin{bmatrix} s(\dot{x}_d\sin\varphi - \dot{y}_d\cos\varphi) & -(\dot{x}_d\cos\varphi + \dot{y}_d\sin\varphi) & 0 \\ -msz_d^{(3)}\sin\varphi & mz_d^{(3)}\cos\varphi & 0 \\ msz_d^{(3)}\cos\varphi & mz_d^{(3)}\sin\varphi & 0 \end{bmatrix} \tag{5.86a}$$

$$\partial^2_{tt}\bar{\boldsymbol{h}}(t,\boldsymbol{p}) = \begin{bmatrix} (x_d - s\cos\varphi)\ddot{x}_d + \dot{x}_d^2 + (y_d - s\sin\varphi)\ddot{y}_d + \dot{y}_d^2 + \dot{z}_d^2 + z_d\ddot{z}_d \\ m\left(x_d^{(4)}z_d + 2\dot{z}_d x_d^{(3)} + \ddot{x}_d\ddot{z}_d - \ddot{x}_d(g+\ddot{z}_d) - (x_d - s\cos\varphi)z_d^{(4)} - 2\dot{x}_d z_d^{(3)}\right) \\ m\left(y_d^{(4)}z_d + 2\dot{z}_d y_d^{(3)} + \ddot{y}_d\ddot{z}_d - \ddot{y}_d(g+\ddot{z}_d) - (y_d - s\sin\varphi)z_d^{(4)} - 2\dot{y}_d z_d^{(3)}\right) \end{bmatrix} \tag{5.86b}$$

As in the case of the planar overhead crane example, it is also possible to get a closed-form analytical solution from the system of equations (5.25a)–(5.25d). The analytical solution derived in this case serves as a reference solution, which is compared to the numerical results in the simulations performed below.

Minimal coordinates

Next the minimally extended index-3 system of equations (5.33a)–(5.33e) will be considered for the rotary crane in terms of minimal coordinates.

To this end, the load position relative to the suspension point is expressed by means of the cable length l and three angles $(\varphi, \theta_1, \theta_2)$, which is shown in Fig. 5.6. That is,

$$\begin{bmatrix} x_d - s\cos\varphi \\ y_d - s\sin\varphi \\ z_d \end{bmatrix} = l\boldsymbol{n}(\varphi, \theta_1, \theta_2) \tag{5.87}$$

Here, the vector $\boldsymbol{n}(\varphi, \theta_1, \theta_2) = \boldsymbol{R}(\varphi, \theta_1, \theta_2)(-\boldsymbol{e}_3)$ is a unit vector (see Fig. 2.6) that points from the suspension point to the load and follows from the canonical base vector $-\boldsymbol{e}_3 = \begin{bmatrix} 0 & 0 & -1 \end{bmatrix}^T$ by applying successive elementary rotations with angles $(\theta_2, \theta_1, \varphi)$ about fixed axes $(-\boldsymbol{e}_2, -\boldsymbol{e}_1, \boldsymbol{e}_3)$ (see also Subsection 2.4.1). This procedure leads to the associated rotation matrix $\boldsymbol{R}(\varphi, \theta_1, \theta_2)$ and eventually yields

$$\boldsymbol{n}(\varphi, \theta_1, \theta_2) = \begin{bmatrix} \sin\theta_2\cos\varphi + \cos\theta_2\sin\theta_1\sin\varphi \\ \sin\theta_2\sin\varphi \quad \cos\theta_2\sin\theta_1\cos\varphi \\ -\cos\theta_2\cos\theta_1 \end{bmatrix} \tag{5.88}$$

Then the coordinate mappings in Equation (5.27) can be written in the form

$$\boldsymbol{\varphi}_1(\boldsymbol{\mu}) = \begin{bmatrix} \varphi \\ s \\ l \end{bmatrix} \quad \text{and} \quad \boldsymbol{\varphi}_2(\boldsymbol{\mu}) = \begin{bmatrix} (s + l\sin\theta_2)\cos\varphi + l\cos\theta_2\sin\theta_1\sin\varphi \\ (s + l\sin\theta_2)\sin\varphi - l\cos\theta_2\sin\theta_1\cos\varphi \\ -l\cos\theta_2\cos\theta_1 \end{bmatrix} \tag{5.89}$$

such that $f = n - m = 5$ minimal coordinates

$$\boldsymbol{\mu} = \begin{bmatrix} \varphi & s & l & \theta_1 & \theta_2 \end{bmatrix}^T \tag{5.90}$$

are used. The same set of coordinates has also been employed in [28]. For the minimal extension procedure of index reduction, the minimal coordinates are split into

$$\boldsymbol{\mu}_1 = \begin{bmatrix} \varphi \\ s \end{bmatrix} \quad \text{and} \quad \boldsymbol{\mu}_2 = \begin{bmatrix} l \\ \theta_1 \\ \theta_2 \end{bmatrix} \tag{5.91}$$

In this case there is

$$D_2\boldsymbol{\varphi}_2(\boldsymbol{\mu}_1,\boldsymbol{\mu}_2) = \begin{bmatrix} \sin\theta_2\cos\varphi + \cos\theta_2\sin\theta_1\sin\varphi & l\cos\theta_2\cos\theta_1\sin\varphi & l\cos\theta_2\cos\varphi - l\sin\theta_2\sin\theta_1\sin\varphi \\ \sin\theta_2\sin\varphi - \cos\theta_2\sin\theta_1\cos\varphi & -l\cos\theta_2\cos\theta_1\cos\varphi & l\cos\theta_2\sin\varphi + l\sin\theta_2\sin\theta_1\cos\varphi \\ -\cos\theta_2\cos\theta_1 & l\cos\theta_2\sin\theta_1 & l\sin\theta_2\cos\theta_1 \end{bmatrix} \tag{5.92}$$

This matrix is nonsingular for realistic parameter values ($l > 0$ and $|\theta_2| < \frac{\pi}{2}$). Accordingly, condition (5.28) is satisfied. The partial derivatives are expressed by

$$D_1\boldsymbol{\varphi}_1(\boldsymbol{\mu}_1,\boldsymbol{\mu}_2) = \begin{bmatrix} 1 & 0 \\ 0 & 1 \\ 0 & 0 \end{bmatrix} \tag{5.93a}$$

$$D_2\boldsymbol{\varphi}_1(\boldsymbol{\mu}_1,\boldsymbol{\mu}_2) = \begin{bmatrix} 0 & 0 & 0 \\ 0 & 0 & 0 \\ 1 & 0 & 0 \end{bmatrix} \tag{5.93b}$$

$$D_1\boldsymbol{\varphi}_2(\boldsymbol{\mu}_1,\boldsymbol{\mu}_2) = \begin{bmatrix} -(s+l\sin\theta_2)\sin\varphi + l\cos\theta_2\sin\theta_1\cos\varphi & \cos\varphi \\ (s+l\sin\theta_2)\cos\varphi + l\cos\theta_2\sin\theta_1\sin\varphi & \sin\varphi \\ 0 & 0 \end{bmatrix} \tag{5.93c}$$

Now, Equation (5.32) gives rise to $\boldsymbol{g}_1 = \boldsymbol{0}$. Furthermore, $\boldsymbol{g}_2(\boldsymbol{\mu},\dot{\boldsymbol{\mu}}_1,\widehat{\boldsymbol{\mu}}_2)$ and $\boldsymbol{h}_\alpha(\boldsymbol{\mu},\dot{\boldsymbol{\mu}}_1,\widehat{\boldsymbol{\mu}}_2)$ can be calculated straightforwardly from Equation (5.32) and (5.34b), respectively. In this connection, it is to be noted that the minimal extension procedure implies the equalities $\hat{l} = \dot{l}$, $\hat{\theta}_1 = \dot{\theta}_1$ and $\hat{\theta}_2 = \dot{\theta}_2$. Eventually, (5.34a)–(5.34c) yields

$$\mathcal{M}_{11}(\boldsymbol{\mu}) = \begin{bmatrix} J_b + m_t s^2 + m\left((s+l\sin\theta_2)^2 + (l\cos\theta_2\sin\theta_1)^2\right) & ml\cos\theta_2\sin\theta_1 \\ ml\cos\theta_2\sin\theta_1 & m+m_t \end{bmatrix} \tag{5.94a}$$

$$\mathcal{M}_{12}(\boldsymbol{\mu}) = \begin{bmatrix} -ms\cos\theta_2\sin\theta_1 & -ml(s+l\sin\theta_2)\cos\theta_2\cos\theta_1 & ml\sin\theta_1(l+s\sin\theta_2) \\ m\sin\theta_2 & 0 & ml\cos\theta_2 \end{bmatrix} \tag{5.94b}$$

$$\mathcal{M}_{21}(\boldsymbol{\mu}) = \mathcal{M}_{12}^T(\boldsymbol{\mu}) \tag{5.94c}$$

$$\mathcal{M}_{22}(\boldsymbol{\mu}) = \begin{bmatrix} m + \frac{J_w}{r_w^2} & 0 & 0 \\ 0 & ml^2\cos^2\theta_2 & 0 \\ 0 & 0 & ml^2 \end{bmatrix} \tag{5.94d}$$

and

$$\mathcal{B}_1^T(\mu) = \begin{bmatrix} 1 & 0 & 0 \\ 0 & 1 & 0 \end{bmatrix}, \quad \mathcal{B}_2^T(\mu) = \begin{bmatrix} 0 & 0 & \frac{1}{r_w} \\ 0 & 0 & 0 \\ 0 & 0 & 0 \end{bmatrix} \tag{5.95}$$

Inverse dynamics simulation

In the numerical simulation the data provided in [28] are used. In particular, the inertia parameters are given by $m = 100\,\text{kg}$, $m_t = 10\,\text{kg}$, $J_w = 0.1\,\text{kg}\cdot\text{m}^2$, $r_w = 0.1\,\text{m}$, and $J_b = 480\,\text{kg}\cdot\text{m}^2$. The servo constraints are used to prescribe a rest-to-rest maneuver of the load specified by

$$\gamma(t) = \gamma_0 + (\gamma_f - \gamma_0)c(t) \tag{5.96}$$

with the initial position

$$\gamma_0 = \begin{bmatrix} 5\,\text{m} \\ 0 \\ -5\,\text{m} \end{bmatrix} \quad \text{at} \quad t_0 = 0 \tag{5.97}$$

and the final position

$$\gamma_f = \begin{bmatrix} -2\,\text{m} \\ 2\,\text{m} \\ -2\,\text{m} \end{bmatrix} \quad \text{at} \quad t_f = 20\,\text{s} \tag{5.98}$$

The reference function $c(t)$ is composed of three phases,

$$c(t) = \begin{cases} c_I(t) & \text{for } \ 0 \leqslant t < 5\,\text{s} \\ c_{II}(t) & \text{for } \ 5\,\text{s} \leqslant t < 15\,\text{s} \\ c_{III}(t) & \text{for } 15\,\text{s} \leqslant t \leqslant 20\,\text{s} \end{cases} \tag{5.99}$$

with each phase

$$c_I(t) = \frac{1}{\tau - \tau_0}\left(-\frac{5t^8}{2\tau_0{}^7} + \frac{10t^7}{\tau_0{}^6} - \frac{14t^6}{2\tau_0{}^5} + \frac{7t^5}{2\tau_0{}^4}\right) \tag{5.100a}$$

$$c_{II}(t) = \frac{1}{\tau - \tau_0}\left(t - \frac{\tau_0}{2}\right) \tag{5.100b}$$

$$c_{III}(t) = 1 + \frac{1}{\tau - \tau_0}\left(-\frac{5(\tau - t)^8}{2\tau_0{}^7} + \frac{10(\tau - t)^7}{\tau_0{}^6} - \frac{14(\tau - t)^6}{2\tau_0{}^5} + \frac{7(\tau - t)^5}{2\tau_0{}^4}\right) \tag{5.100c}$$

Using the minimal coordinates (5.90), the initial configuration of the rotary crane at $t_0 = 0$ is defined by

$$\boldsymbol{\mu}_0 = \begin{bmatrix} 0 & 5\,\mathrm{m} & 5\,\mathrm{m} & 0 & 0 \end{bmatrix}^T \tag{5.101}$$

The motion of the crane is starting at rest such that $\dot{\boldsymbol{\mu}}_0 = \mathbf{0}$.

The simulation results for different time step sizes are depicted in Fig. 5.7 and 5.8. In each diagram, the numerical solution (NUM) is compared to the analytical reference solution (REF). It can be observed that the numerical solution converges to the analytical reference solution when the time step size is reduced. Both formulations in terms of redundant and minimal coordinates yield practically the same results. Similar observations can be made for the implementation of the projection method due to [31]. The motion of the rotary crane is illustrated in Fig. 5.9 with some snapshots at consecutive points in time.

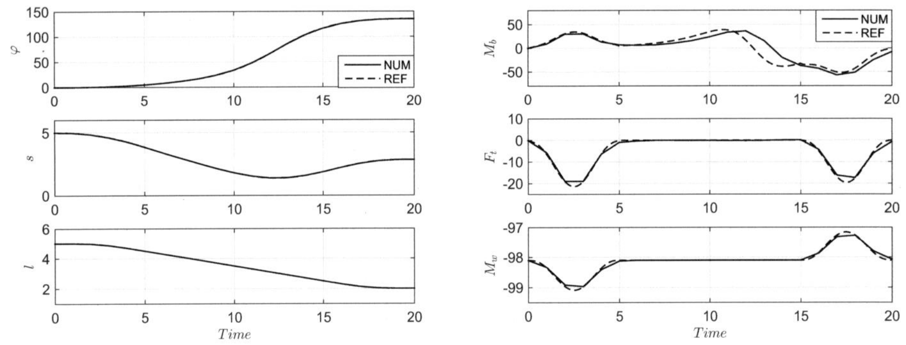

Figure 5.7: Rotary crane: Comparison between the numerical results (NUM) obtained with $\Delta t = 1\,\mathrm{s}$ and the analytical reference solution (REF).

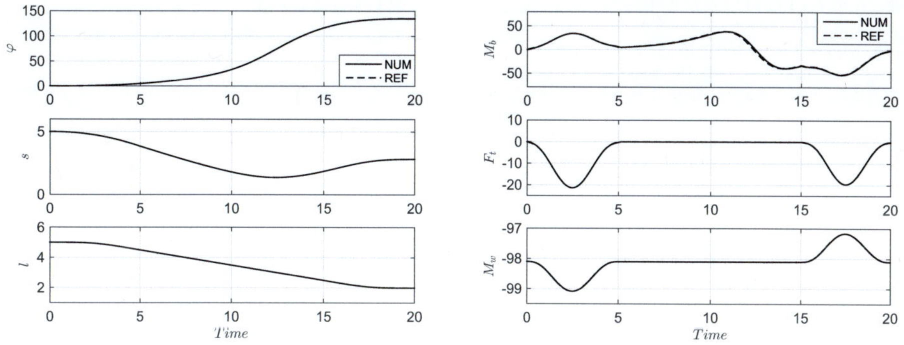

Figure 5.8: Rotary crane: Comparison between the numerical results (NUM) obtained with $\Delta t = 0.1\,\mathrm{s}$ and the analytical reference solution (REF).

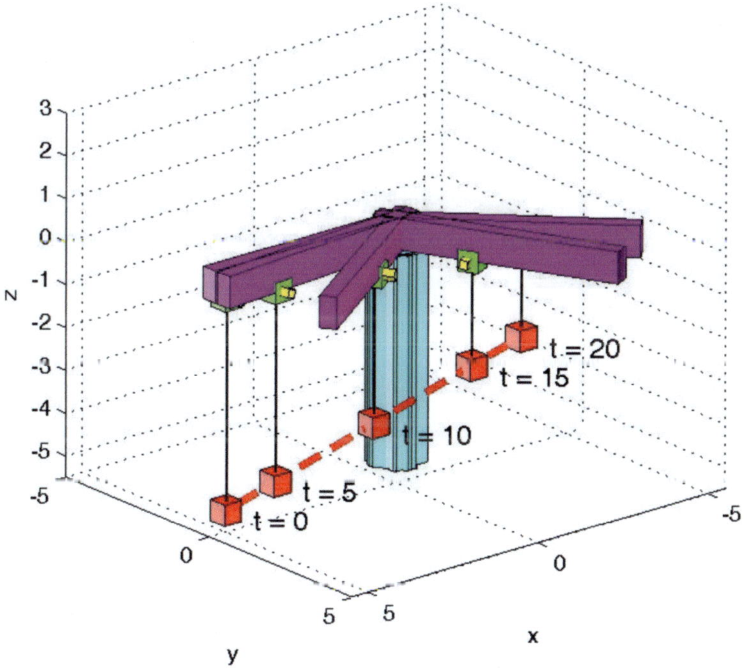

Figure 5.9: Rotary crane: Snapshots at specific points in time.

5.4 Redundant coordinates formulation

In Subsection 5.1.2, the index reduction by minimal extention procedure has been discussed in the dependent coordinates formulation of cranes. This index reduction appoach is based on the introduction of new algebraic variables along with the enlargement of the DAEs by appending time derivatives of the constraints.

It has been shown that index reduction by minimal extension can be applied very efficiently by exploiting the specific structure provided by underactuated mechanical systems. In this connection, either minimal or redundant coordinates can be used.

It is verified that the index reduction by minimal extension approach is a viable alternative to the projection method. It has also been shown that in a first step the minimal extension approach can be used to lower the index of the DAEs from five to three and in a second step the index can even be reduced to one.

The next goal of the present work is to extend the applicability of the index reduction approach to mechanical models of underactuated systems that rely on arbitrarily selected redundant coordinates. Specifically, in contrast to the above sections in this chapter, the number of holonomic constraints is not limited. Consequently, general crane formulations such as those developed in [58] can now be included into the present index reduction approach. Similarly, other rotationless formulations of multibody dynamics such as natural coordinates or Cosserat-type descriptions in terms of directors (including rigid bodies and nonlinear beams and shells) typically yield a large number of holonomic constraints. These formulations are now embraced as well by the newly developed index reduction method. In the following part the main focus is placed on the inverse dynamics of a family of crane models that are known to belong to the class of differentially flat systems.

5.4.1 Inverse dynamics of underactuated mechanical systems

At first a general formulation of mechanical systems subjected to both holonomic and servo constraints will be introduced. In particular, the equations of motion of the following form

$$\begin{bmatrix} M_1(p) & 0 \\ 0 & M_2 \end{bmatrix} \begin{bmatrix} \ddot{p} \\ \ddot{x} \end{bmatrix} = \begin{bmatrix} f_1(p,\dot{p}) \\ f_2(x,\dot{x}) \end{bmatrix} + \begin{bmatrix} B_1^T(p) \\ 0 \end{bmatrix} u - G^T(p,x)\lambda \tag{5.102a}$$

$$0 = h(p) \tag{5.102b}$$

$$0 = \Phi(p,x) \tag{5.102c}$$

$$x = \gamma \tag{5.102d}$$

are considered. The first row block in (5.102a) corresponds to the robot (or input) subsystem with coordinates $p \in \mathbb{R}^{n-a}$, whereas the second row block in (5.102a) corresponds to the output subsystem with coordinates $x \in \mathbb{R}^a$. The n redundant coordinates

$$q = \begin{bmatrix} p \\ x \end{bmatrix} \tag{5.103}$$

are subject to the holonomic constraints (5.102b) and (5.102c), with associated constraint functions $h \in \mathbb{R}^{m_1}$ and $\Phi \in \mathbb{R}^{m_2}$. The total number of holonomic constraints is denoted by $m = m_1 + m_2$. Note that the constraint function h does not depend on the output variables x. The Jacobian of the holonomic constraints assumes the form

$$G(p,x) = \begin{bmatrix} H_1(p) & 0 \\ G_1(p,x) & G_2(p,x) \end{bmatrix} = \begin{bmatrix} \partial_p h(p) & 0 \\ \partial_p \Phi(p,x) & \partial_x \Phi(p,x) \end{bmatrix} \in \mathbb{R}^{m,n} \tag{5.104}$$

The Lagrange multipliers associated to the m holonomic constraints are contained in the vector

$$\lambda = \begin{bmatrix} \lambda_1 \\ \lambda_2 \end{bmatrix} \in \mathbb{R}^m, \quad \text{with} \quad \lambda_1 \in \mathbb{R}^{m_1}, \lambda_2 \in \mathbb{R}^{m_2} \tag{5.105}$$

Due to the presence of holonomic constraints, the configuration space of the constrained mechanical system under consideration is defined by

$$Q = \{q \in \mathbb{R}^n | h(p) = 0, \Phi(p,x) = 0\} \tag{5.106}$$

It is assumed that the constraints are independent. Consequently, the constraint Jacobian G has full row rank and the discrete mechanical system under consideration has $n - m$ degrees of freedom.

The servo constraints (5.102d) specify the desired trajectory of the load via the prescribed function $\gamma \colon \mathbb{I} \rightarrow \mathbb{R}^a$, where $\mathbb{I} = [t_0, t_f]$ is the time interval of interest.

Note that the attention is focused on underactuated mechanical systems in which the number of controls is fewer than the number of degrees of freedom, that is, $a < n - m$.

The control inputs $\boldsymbol{u} \in \mathbb{R}^a$ regulate the control forces acting on the robot subsystem. In this connection, the matrix $\boldsymbol{B}_1 \in \mathbb{R}^{a,n-a}$ denotes the input transformation matrix. Besides the constraint and control forces, additional forces acting on the system are contained in the conjugate force vectors $\boldsymbol{f}_1 \in \mathbb{R}^{n-a}$ and $\boldsymbol{f}_2 \in \mathbb{R}^a$. Similarly, the mass matrix is split into the submatrices $\boldsymbol{M}_1 \in \mathbb{R}^{n-a,n-a}$ and $\boldsymbol{M}_2 \in \mathbb{R}^{a,a}$.

Due to the presence of servo constraints, the index of the DAEs (5.102a)–(5.102d) often exceeds 3. This has been proved in the examples before. For example, the application of servo constraints to (differentially flat) crane systems typically yields an index of 5. Consequently, prior to the application of a numerical integrator the index of the DAEs should be lowered. For that purpose, following the treatment in the previous sections (see also [2]), the index reduction by minimal extension can be applied to the DAEs (5.102a)–(5.102d) as well.

At this point it should be emphasized that in the above formulation (5.102a)–(5.102d) the number of holonomic constraints, m, is just restricted by $m < n$. This facilitates the arbitrary selection of redundant coordinates best suited for the description and numerical simulation of the specific inverse dynamics problem at hand. This is in contrast to the case treated in the previous sections in this chapter (see also [2]), where $m \leqslant a$ has been assumed.

5.4.2 Index reduction by minimal extension procedure

Guided by the mininal extension procedure [2], the system of DAEs (5.102a)–(5.102d) is enlarged by appending the first and second time derivative of the servo constraints.

To maintain a square system, additional dummy derivatives $\widehat{\boldsymbol{x}} := \dot{\boldsymbol{x}}$ and $\widetilde{\boldsymbol{x}} := \ddot{\boldsymbol{x}}$ are introduced, and they replace the corresponding derivatives of the outputs $\boldsymbol{x}$ in the following square system of equations.

Accordingly, the minimally extended system reads

$$\begin{bmatrix} M_1(p) & 0 \\ 0 & M_2 \end{bmatrix} \begin{bmatrix} \dot{p} \\ \widetilde{x} \end{bmatrix} = \begin{bmatrix} f_1(p, \dot{p}) \\ f_2(x, \widehat{x}) \end{bmatrix} + \begin{bmatrix} B_1^T(p) \\ 0 \end{bmatrix} u - \begin{bmatrix} H_1^T(p) & G_1^T(p, x) \\ 0 & G_2^T(p, x) \end{bmatrix} \begin{bmatrix} \lambda_1 \\ \lambda_2 \end{bmatrix} \tag{5.107a}$$

$$0 = h(p) \tag{5.107b}$$

$$0 = \Phi(p, x) \tag{5.107c}$$

$$x = \gamma \tag{5.107d}$$

$$\widehat{x} = \dot{\gamma} \tag{5.107e}$$

$$\widetilde{x} = \ddot{\gamma} \tag{5.107f}$$

Next it will be shown that - provided certain assumptions apply - the minimally extended system (5.107a)–(5.107f) has index 3. Typical applications are differentially flat crane models[4] where the index equals 5 in the original form. Previously the special case $m \leqslant a$ and $M_1(p)$ non-singular has been shown. Here it is allowed that there are more holonomic constraints than servo constraints.

To guarantee the index-3 property of system of equations (5.107a)–(5.107f), the following two assumptions are stated. The first assumption ensures, amongst others, that the number of holonomic constraints depending on x is bounded by the dimension of x, namely a.

Assumption 2. *The block $G_2^T(p, \gamma)$ of the Jacobian G is of full rank and $m_2 \leqslant a$. Furthermore, the dimensions satisfy*

$$2a + m_1 \leqslant n \quad \textit{and} \quad a < n - m$$

The last inequality ensures that the system is underactuated. Note that the two previous assumptions already imply $a \leqslant n - m$.

Within the proof of Theorem 5.2 below, the equations will be reduced and rewritten such that the resulting system has the typical multibody structure and thus, is of index 3. This requires a certain matrix to be invertible, which will be summarized in the following assumption.

[4] Some advanced crane examples are given in Chapter 6.

Assumption 3. *If Assumption 2 is satisfied, then there exists a matrix* $\boldsymbol{P}_2 \in \mathbb{R}^{a,a-m_2}$*, whose columns span the null space of* $\boldsymbol{G}_2$*, i.e.,* $\boldsymbol{G}_2(\boldsymbol{p},\boldsymbol{\gamma})\boldsymbol{P}_2(\boldsymbol{p},\boldsymbol{\gamma}) = \boldsymbol{0}$*. Then* $\boldsymbol{z} \in \mathbb{R}^{a-m_2}$ *is defined by the equation*

$$\boldsymbol{z}(\boldsymbol{p},\boldsymbol{\gamma}) := \boldsymbol{P}_2^T(\boldsymbol{p},\boldsymbol{\gamma})\left(\boldsymbol{f}_2(\boldsymbol{\gamma},\dot{\boldsymbol{\gamma}}) - \boldsymbol{M}_2\ddot{\boldsymbol{\gamma}}\right) \tag{5.108}$$

and its derivative with respect to $\boldsymbol{p}$ *is denoted by* $\boldsymbol{Z}_1(\boldsymbol{p},\boldsymbol{\gamma}) = \partial_{\boldsymbol{p}}\boldsymbol{z}(\boldsymbol{p},\boldsymbol{\gamma})$*. After that, with* $\boldsymbol{G}_1$ *and* $\boldsymbol{H}_1$ *from Equation* (5.104)*, it is assumed that the matrix*

$$\begin{bmatrix} \boldsymbol{M}_1(\boldsymbol{p}) & -\boldsymbol{B}_1^T(\boldsymbol{p}) & \boldsymbol{H}_1^T(\boldsymbol{p}) \\ \boldsymbol{Z}_1(\boldsymbol{p},\boldsymbol{\gamma}) & \boldsymbol{0} & \boldsymbol{0} \\ \boldsymbol{G}_1(\boldsymbol{p},\boldsymbol{\gamma}) & \boldsymbol{0} & \boldsymbol{0} \\ \boldsymbol{H}_1(\boldsymbol{p}) & \boldsymbol{0} & \boldsymbol{0} \end{bmatrix} \in \mathbb{R}^{n+m_1,n+m_1} \tag{5.109}$$

is invertible. Note that the matrix $\boldsymbol{M}_1(\boldsymbol{p})$ *itself is not asked to be invertible.*

With the two assumptions in hand, the following theorem can be formulated.

Theorem 5.2. *Given the Assumptions 2 and 3, the extended system* (5.107a)–(5.107f) *is a set of index-3 DAEs.*

Proof. As mentioned before, the idea of proof is to reduce the DAEs (5.107a)–(5.107f) to a system which has a similar structure as a constrained multibody system.

Since the variables $\boldsymbol{x}$, $\hat{\boldsymbol{x}}$ and $\tilde{\boldsymbol{x}}$ are directly given by the prescribed trajectory $\boldsymbol{\gamma}$ and its derivatives, they may be eliminated from the system equations. Consider the second part of Equation (5.107a), namely

$$\boldsymbol{M}_2\ddot{\boldsymbol{\gamma}} = \boldsymbol{f}_2(\boldsymbol{\gamma},\dot{\boldsymbol{\gamma}}) - \boldsymbol{G}_2^T(\boldsymbol{p},\boldsymbol{\gamma})\boldsymbol{\lambda}_2.$$

Then the full rank property of $\boldsymbol{G}_2$ together with $m_2 \leqslant a$ from Assumption 2 implies that $\boldsymbol{\lambda}_2$ can be obtained by

$$\boldsymbol{\lambda}_2 = \left(\boldsymbol{G}_2(\boldsymbol{p},\boldsymbol{\gamma})\boldsymbol{G}_2^T(\boldsymbol{p},\boldsymbol{\gamma})\right)^{-1}\boldsymbol{G}_2(\boldsymbol{p},\boldsymbol{\gamma})\left(\boldsymbol{f}_2(\boldsymbol{\gamma},\dot{\boldsymbol{\gamma}}) - \boldsymbol{M}_2\ddot{\boldsymbol{\gamma}}\right).$$

With the matrix $\boldsymbol{P}_2 \in \mathbb{R}^{a,a-m_2}$ from Assumption 3, which spans the null space of $\boldsymbol{G}_2$, $\boldsymbol{z}(\boldsymbol{p},\boldsymbol{\gamma}) \in \mathbb{R}^{a-m_2}$ can be defined according to Equation (5.108). Note that this defines an algebraic constraint, that is

$$\boldsymbol{z}(\boldsymbol{p},\boldsymbol{\gamma}) = \boldsymbol{0}.$$

In summary, the remaining variables need to satisfy the system

$$M_1(p)\ddot{p} = \overline{f}_1(p,\dot{p}) + B_1^T(p)u - H_1^T(p)\lambda_1 \tag{5.110a}$$
$$0 = z(p,\gamma) \tag{5.110b}$$
$$0 = \Phi(p,\gamma) \tag{5.110c}$$
$$0 = h(p) \tag{5.110d}$$

with the term

$$\overline{f}_1(p,\dot{p}) := f_1(p,\dot{p}) - G_1^T(p,\gamma)\lambda_2 = f_1 - G_1^T\left(G_2G_2^T\right)^{-1}G_2(f_2 - M_2\ddot{\gamma})$$

The resulting system of equations (5.110a)–(5.110d) consists of $n-a$ dynamic equations and $m_2+m_1+(a-m_2) = a+m_1$ constraints. After replacing the three constraints by their second derivatives, the system of equations (5.110a)–(5.110d) can be written in the form

$$\hat{M}\begin{bmatrix}\ddot{p}\\ u\\ \lambda_1\end{bmatrix} = \begin{bmatrix}\overline{f}_1(p,\dot{p})\\ z_2(p,\dot{p},\gamma,\dot{\gamma},\ddot{\gamma})\\ \Phi_2(p,\dot{p},\gamma,\dot{\gamma},\ddot{\gamma})\\ h_2(p,\dot{p})\end{bmatrix}$$

Therein, $\hat{M}$ denotes the matrix in (5.109) and z_2, Φ_2, and h_2 are the functions which include the remaining terms of the differentiation. Since $\hat{M}$ is non-singular by Assumption 3, a multiplication by its inverse matrix from the left yields an ODE for p and algebraic equations for u and λ_1. Since only two differentiations are necessary, the system (5.110a)–(5.110d) and thus, also system (5.107a)–(5.107f) need to be (at most) of index 3. □

Reduction of the number of redundant coordinates

Next the present formulation will be connected to the formulation described in Subsection 5.1.2. To this end, the holonomic constraints (5.107b) are eliminated by reducing the number of redundant coordinates from n to $\overline{n} = n - m_1$. This is possible if a mapping $\varphi : \mathbb{R}^{\overline{n}-a} \to \mathbb{R}^{n-a}$ can be found such that

$$p = \varphi(\overline{p}) \tag{5.111}$$

where $\overline{p} \in \mathbb{R}^{\overline{n}-a}$ denotes the new redundant coordinates. The mapping (5.111) has to satisfy the constraints (5.107b) identically for arbitrary $\overline{p} \in \mathbb{R}^{\overline{n}-a}$. Consequently,

$$h(\varphi(\overline{p})) = 0 \quad \text{and} \quad H_1(p)D\varphi(\overline{p}) = 0 \tag{5.112}$$

for $p = \varphi(\overline{p})$. Premultiplying the first row block in Equation (5.107a) by $D\varphi^T(\overline{p})$ and taking into account Equation (5.111) and (5.112) yield the size-reduced DAEs

$$\begin{bmatrix} \overline{M}_1(\overline{p}) & 0 \\ 0 & M_2 \end{bmatrix} \begin{bmatrix} \ddot{\overline{p}} \\ \breve{x} \end{bmatrix} = \begin{bmatrix} \overline{f}_1(\overline{p}, \dot{\overline{p}}) \\ f_2(x, \widehat{x}) \end{bmatrix} + \begin{bmatrix} \overline{B}_1^T(\overline{p}) \\ 0 \end{bmatrix} u - \begin{bmatrix} \overline{G}_1^T(\overline{p}, x) \\ \overline{G}_2^T(\overline{p}, x) \end{bmatrix} \lambda_2 \tag{5.113a}$$

$$0 = \overline{\Phi}(\overline{p}, x) \tag{5.113b}$$

$$x = \gamma \tag{5.113c}$$

$$\widehat{x} = \dot{\gamma} \tag{5.113d}$$

$$\breve{x} = \ddot{\gamma} \tag{5.113e}$$

where

$$\overline{M}_1(\overline{p}) = D\varphi^T(\overline{p}) M_1(p) D\varphi(\overline{p}) \tag{5.114a}$$

$$\overline{f}_1(\overline{p}, \dot{\overline{p}}) = D\varphi^T(\overline{p}) \left(f_1(p, \dot{p}) - M_1(\overline{p}) \frac{d}{dt} (D\varphi(\overline{p})) \dot{\overline{p}} \right) \tag{5.114b}$$

$$\overline{B}_1^T(\overline{p}) = D\varphi^T(\overline{p}) B_1^T(p) \tag{5.114c}$$

$$\overline{\Phi}(\overline{p}, x) = \Phi(p, x) \tag{5.114d}$$

$$\overline{G}_1(\overline{p}, x) = G_1(p, x) D\varphi(\overline{p}) \tag{5.114e}$$

$$\overline{G}_2(\overline{p}, x) = \partial_x \overline{\Phi}(\overline{p}, x) \tag{5.114f}$$

for $p = \varphi(\overline{p})$. In this way, the number of redundant coordinates is reduced by m_1 such that the remaining coordinates are given by the $\overline{n}$-dimensional configuration vector

$$\overline{q} = \begin{bmatrix} \overline{p} \\ x \end{bmatrix} \tag{5.115}$$

Note that Equation (5.113b) contains the m_2 remaining holonomic constraints with associated Lagrange multipliers $\lambda_2 \in \mathbb{R}^{m_2}$ in Equation (5.113a). The configuration space of the constrained mechanical system under consideration can now be expressed in the form

$$Q = \{\overline{q} \in \mathbb{R}^{\overline{n}} | \overline{\Phi}(\overline{p}, x) = 0\} \tag{5.116}$$

The DAEs (5.113a)–(5.113e) lie at the heart of the work in previous sections of this chapter. It has been shown that under certain conditions (e.g. $m_2 \leqslant a$ and $\overline{M}_1(\overline{p})$ non-singular) the minimally extended DAEs (5.113a)–(5.113e) attain an index reduction by two. In the case of differentially flat crane models, the original DAEs have index 5, whereas the index-reduced DAEs (5.113a)–(5.113e) (and, correspondingly, the DAEs (5.107a)–(5.107f) as well) have index 3. It has also been shown that a second application of index reduction by minimal extension can achieve a reduction to index-1 DAEs.

5.4.3 Numerical discretization

For the specific inverse dynamics problems, which will be dealt with in the next subsection, the proposed index reduction approach yields DAEs (5.107a)–(5.107f) with index 3. Due to the semi-explicit form of the DAEs (5.107a)–(5.107f), the simple backward Euler discretization can be expected to work well (see Ascher and Petzold [6, Sec. 10.1.1]). The DAEs (5.107a)–(5.107f) can be recast in the form

$$M_1(p)\ddot{p} = f_1(p,\dot{p}) + B_1^T(p)u - H_1^T(p)\lambda_1 - G_1^T(p,\gamma)\lambda_2 \tag{5.117a}$$
$$0 = M_2\ddot{\gamma} - f_2(\gamma,\dot{\gamma}) + G_2^T(p,\gamma)\lambda_2 \tag{5.117b}$$
$$0 = \Phi(p,\gamma) \tag{5.117c}$$
$$0 = h(p) \tag{5.117d}$$

The DAEs (5.117a)–(5.117d) provide $n-a$ differential equations (5.117a) along with $a+m$ algebraic equations (5.117b) through (5.117d) for the determination of $p \in \mathbb{R}^{n-a}$, $u \in \mathbb{R}^a$, and $\lambda \in \mathbb{R}^m$. Application of the backward Euler method yields

$$p_{n+1} - p_n = \Delta t v_{n+1} \tag{5.118a}$$
$$M_1(p_{n+1})\frac{(v_{n+1} - v_n)}{\Delta t} = f_1(p_{n+1}, v_{n+1}) + B_1^T(p_{n+1})u_{n+1} - H_1^T(p_{n+1})\lambda_{1_{n+1}} - G_1^T(p_{n+1}, \gamma(t_{n+1}))\lambda_{2_{n+1}} \tag{5.118b}$$
$$0 = M_2\ddot{\gamma}(t_{n+1}) - f_2(\gamma(t_{n+1}), \dot{\gamma}(t_{n+1})) + G_2^T(p_{n+1}, \gamma(t_{n+1}))\lambda_{2_{n+1}} \tag{5.118c}$$
$$0 = \Phi(p_{n+1}, \gamma(t_{n+1})) \tag{5.118d}$$
$$0 = h(p_{n+1}) \tag{5.118e}$$

In a typical time step of size $\Delta t = t_{n+1} - t_n$ approximations $(\bullet)_{n+1}$ to $(\bullet)(t_{n+1})$ need to be found if the corresponding quantities $(\bullet)_n$ are given as the result of the previous step. For the initial time step, the consistent initial values $\boldsymbol{p}_0$ and $\boldsymbol{v}_0$ are required and they have to satisfy $\boldsymbol{\Phi}(\boldsymbol{p}_0, \gamma(t_0)) = \boldsymbol{0}$ and $\boldsymbol{h}(\boldsymbol{p}_0) = \boldsymbol{0}$ along with

$$\partial_{\boldsymbol{p}}\boldsymbol{\Phi}(\boldsymbol{p}_0, \gamma(t_0))\boldsymbol{v}_0 + \partial_{\boldsymbol{x}}\boldsymbol{\Phi}(\boldsymbol{p}_0, \gamma(t_0))\dot{\gamma}(t_0) = \boldsymbol{0} \tag{5.119a}$$

$$\partial_{\boldsymbol{p}}\boldsymbol{h}(\boldsymbol{p}_0)\boldsymbol{v}_0 = \boldsymbol{0} \tag{5.119b}$$

The scheme (5.118a)–(5.118e) provides $2n + m - a$ algebraic equations for the determination of $\boldsymbol{p}_{n+1}, \boldsymbol{v}_{n+1} \in \mathbb{R}^{n-a}$, $\boldsymbol{u}_{n+1} \in \mathbb{R}^a$, and $\boldsymbol{\lambda}_{n+1} \in \mathbb{R}^m$.

5.4.4 Sample application: Three-dimensional rotary crane

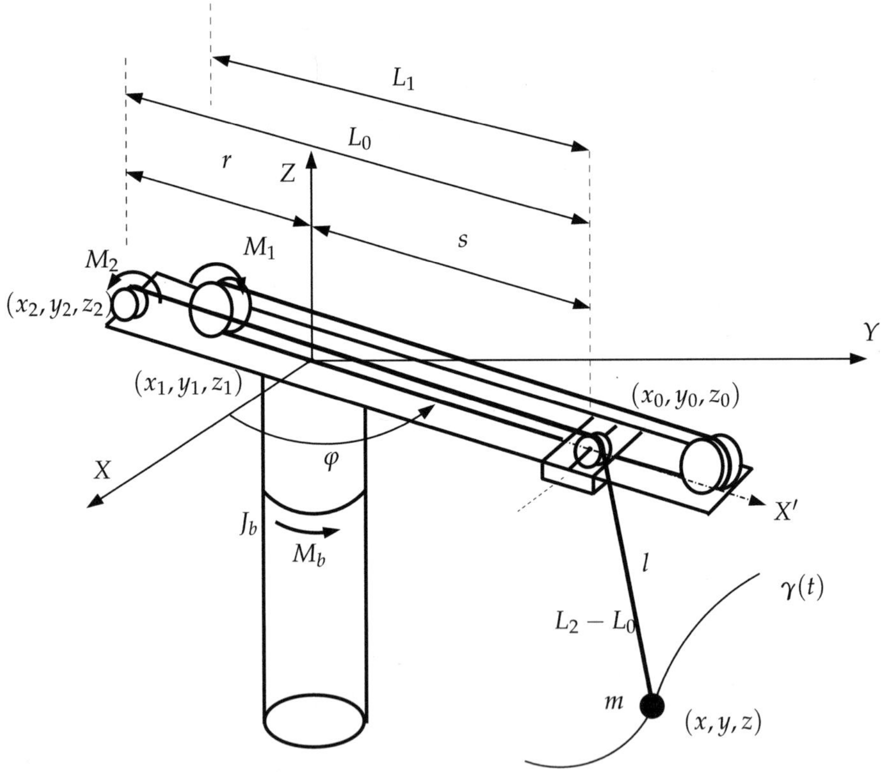

Figure 5.10: The three-dimensional rotary crane model in terms of $n = 10$ redundant coordinates.

The present approach will be demonstrated with the inverse dynamics simulation of a specific three-dimensional rotary crane. Previously, in Section 5.3.2 the rotary crane under consideration has been formulated in terms of $n = 6$ redundant coordinates (see also [2, 31]) or 5 minimal coordinates (see also [2, 28]). Alternatively, the much more general framework [58] is now used for the modeling of cranes. For this, $n = 10$ redundant coordinates are used and they are subject to $m = 5$ holonomic constraints. The enlarged set of redundant crane coordinates, as shown in Fig. 5.10, is given by

$$\boldsymbol{p} = \begin{bmatrix} x_2 & y_2 & x_0 & y_0 & L_1 & L_2 & L_0 \end{bmatrix}^T \tag{5.120}$$

and

$$\boldsymbol{x} = \begin{bmatrix} x & y & z \end{bmatrix}^T \tag{5.121}$$

Similarly, the last equation specifies the load (mass m) coordinates relative to a Cartesian inertial frame. The load is connected to the hoisting winch 2 (Cartesian coordinates $x_2, y_2, z_2 = 0$, actuating torque M_2, radius r_2, moment of inertia J_2) via a cable of length L_2. The position of the trolley (Cartesian coordinates $x_0, y_0, z_0 = 0$, mass m_0) on the girder bridge relative to the hoisting winch 2 is given by L_0. The trolley contains a pulley (radius r_w, moment of inertia J_w) and is moved along the girder bridge under the action of a second winch 1 (Cartesian coordinates $x_1, y_1, z_1 = 0$, actuating torque M_1, radius r_1, moment of inertia J_1) whose position on the girder bridge relative to the hoisting winch 2 is fixed by the parameter $\alpha = \frac{1}{2}$. The distance between winch 1 and the trolley is given by L_1. The holonomic constraints $\boldsymbol{h}(\boldsymbol{p}) = \boldsymbol{0}$ are given by

$$\boldsymbol{h}(\boldsymbol{p}) = \begin{bmatrix} \frac{1}{2}((x_0 - \alpha x_2)^2 + (y_0 - \alpha y_2)^2 - L_1^2) \\ \frac{1}{2}((x_0 - x_2)^2 + (y_0 - y_2)^2 - L_0^2) \\ \frac{1}{2}(x_2^2 + y_2^2 - r^2) \\ x_2 y_0 - x_0 y_2 \end{bmatrix} \tag{5.122}$$

Accordingly, $m_1 = 4$. Note that the first two constraints link the coordinates L_1 and L_0 to the position of the trolley and, respectively, winch 1 and winch 2. Moreover, the third constraint links the parameter r to the position of winch 2, and the fourth constraint confines the relative motion of the trolley to the longitudinal direction along the girder bridge. The last holonomic constraint $\boldsymbol{\Phi}(\boldsymbol{p}, \boldsymbol{x}) = \boldsymbol{0}$ is specified by

$$\Phi(\boldsymbol{p}, \boldsymbol{x}) = \frac{1}{2}((x - x_0)^2 + (y - y_0)^2 + z^2 - (L_2 - L_0)^2) \tag{5.123}$$

and connects the load coordinates with the robot (or crane) coordinates. Accordingly,

$m_2 = 1$. The total kinetic energy of the mechanical system under consideration assumes the form

$$T = \frac{1}{2}\dot{\boldsymbol{p}} \cdot \boldsymbol{M}_1 \dot{\boldsymbol{p}} + \frac{1}{2}\dot{\boldsymbol{x}} \cdot \boldsymbol{M}_2 \dot{\boldsymbol{x}} \tag{5.124}$$

in which the mass matrices corresponding to the robot coordinates and the load coordinates are given by

$$\boldsymbol{M}_1 = \begin{bmatrix} M & 0 & 0 & 0 & 0 & 0 & 0 \\ 0 & M & 0 & 0 & 0 & 0 & 0 \\ 0 & 0 & m_0 & 0 & 0 & 0 & 0 \\ 0 & 0 & 0 & m_0 & 0 & 0 & 0 \\ 0 & 0 & 0 & 0 & \frac{J_1}{r_1^2} + \frac{J_w}{r_w^2} & -\frac{J_w}{r_w^2} & 0 \\ 0 & 0 & 0 & 0 & -\frac{J_w}{r_w^2} & \frac{J_2}{r_2^2} + \frac{J_w}{r_w^2} & 0 \\ 0 & 0 & 0 & 0 & 0 & 0 & 0 \end{bmatrix}, \quad \boldsymbol{M}_2 = \begin{bmatrix} m & 0 & 0 \\ 0 & m & 0 \\ 0 & 0 & m \end{bmatrix} \tag{5.125}$$

Here, the mass M is connected to the moment of inertia of the girder bridge relative to the Z-axis, J_b, via $M = \frac{J_b}{r^2}$. Further quantities needed in system of equations (5.102a)–(5.102d) are given by

$$\boldsymbol{B}_1^T = \begin{bmatrix} 0 & 0 & -\frac{y_2}{r^2} \\ 0 & 0 & \frac{x_2}{r^2} \\ 0 & 0 & 0 \\ 0 & 0 & 0 \\ \frac{1}{r_1} & 0 & 0 \\ 0 & \frac{1}{r_2} & 0 \\ 0 & 0 & 0 \end{bmatrix}, \quad \boldsymbol{f}_1 = \begin{bmatrix} 0 \\ 0 \\ 0 \\ 0 \\ 0 \\ 0 \\ 0 \end{bmatrix}, \quad \boldsymbol{f}_2 = \begin{bmatrix} 0 \\ 0 \\ -mg \end{bmatrix}, \quad \boldsymbol{u} = \begin{bmatrix} M_1 \\ M_2 \\ M_b \end{bmatrix} \tag{5.126}$$

and

$$\boldsymbol{H}_1^T = \begin{bmatrix} \alpha(x_0 - \alpha x_2) & x_0 - x_2 & -x_2 & -y_0 \\ \alpha(y_0 - \alpha_1 y_2) & y_0 - y_2 & -y_2 & x_0 \\ \alpha x_2 - x_0 & x_2 - x_0 & 0 & y_2 \\ \alpha y_2 - y_0 & y_2 - y_0 & 0 & -x_2 \\ L_1 & 0 & 0 & 0 \\ 0 & 0 & 0 & 0 \\ 0 & L_0 & 0 & 0 \end{bmatrix}, \quad \boldsymbol{G}_1^T = \begin{bmatrix} 0 \\ 0 \\ x - x_0 \\ y - y_0 \\ 0 \\ L_2 - L_0 \\ L_0 - L_2 \end{bmatrix}, \quad \boldsymbol{G}_2^T = \begin{bmatrix} x_0 - x \\ y_0 - y \\ -z \end{bmatrix} \tag{5.127}$$

Note that there are $a = 3$ control inputs given by the two winch torques M_1, M_2, along with the torque M_b acting about the Z-axis of the rotary crane.

Check of the assumptions

In order to show that the three-dimensional rotary crane with redundant variables fits into the given framework, it is necessary to check whether Assumptions 2 and 3 are satisfied.

Clearly G_2^T is of full rank if either $x \neq x_0$, $y \neq y_0$, or $z \neq 0$. Note that this is a reasonable assumption since otherwise the position of the trolley would be equal to the position of the load. Furthermore, the dimensions satisfy $n = 10$, $m_1 = 4$, $m_2 = 1$, and $a = 3$ such that

$$1 = m_2 \leqslant a = 3, \quad 10 = 2a + m_1 \leqslant n = 10, \quad 3 = a < n - m = 5 \tag{5.128}$$

For the second assumption, the matrix $P_2 \in \mathbb{R}^{a,a-m_2}$ is needed, which spans the null space of G_2. Depending on the case whether $x \neq x_0$, $y \neq y_0$, or $z \neq 0$, the projection P_2 may be given by

$$P_2 = \begin{bmatrix} y_0 - y & z \\ x - x_0 & 0 \\ 0 & x - x_0 \end{bmatrix}, \quad P_2 = \begin{bmatrix} y_0 - y & 0 \\ x - x_0 & z \\ 0 & y_0 - y \end{bmatrix}, \text{ or } \quad P_2 = \begin{bmatrix} z & 0 \\ 0 & z \\ x_0 - x & y_0 - y \end{bmatrix} \tag{5.129}$$

Without loss of generality, it is assumed in the sequel that $z \neq 0$ which leads to

$$z = P_2^T(p,\gamma)\left(f_2(\gamma,\dot{\gamma}) - M_2\ddot{\gamma}\right) = -mg \begin{bmatrix} x_0 - x \\ y_0 - y \end{bmatrix} - m \begin{bmatrix} z\ddot{\gamma}_1 + (x_0 - x)\ddot{\gamma}_3 \\ z\ddot{\gamma}_2 + (y_0 - y)\ddot{\gamma}_3 \end{bmatrix} \tag{5.130}$$

and thus,

$$Z_1(p,\gamma) = \partial_p z(p,\gamma) = \begin{bmatrix} 0 & 0 & -m(g+\ddot{\gamma}_3) & 0 & 0 & 0 & 0 \\ 0 & 0 & 0 & -m(g+\ddot{\gamma}_3) & 0 & 0 & 0 \end{bmatrix} \tag{5.131}$$

Since in this special case the matrices

$$\begin{bmatrix} -B_1^T(p) & H_1^T(p) \end{bmatrix} \quad \text{and} \quad \begin{bmatrix} Z_1^T(p,\gamma) & G_1^T(p,\gamma) & H_1^T(p) \end{bmatrix} \tag{5.132}$$

are square, it is sufficient to show the invertibility of the two matrices in order to prove that the matrix in (5.109) is invertible and thus Assumption 3 is satisfied. A close look at the matrices then shows that minimal extension reduces the system equations of the three-dimensional rotary crane to index-3 DAEs if the following conditions are satisfied:

$$z, L_0, L_1, L_2 \neq 0, \quad L_0 \neq L_2, \quad g \neq -\ddot{\gamma}_3, \quad x_0 x_2 + y_0 y_2 \neq 0, \quad x_2^2 + y_2^2 \neq 0 \tag{5.133}$$

Reduction of the number of redundant coordinates

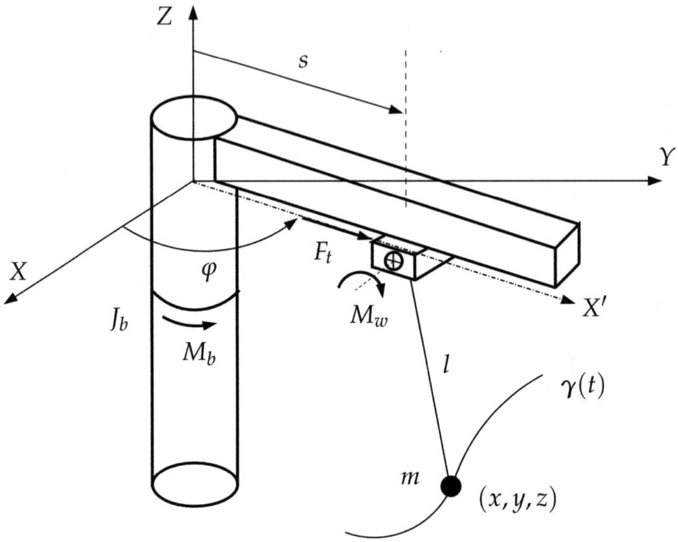

Figure 5.11: The three-dimensional rotary crane model in terms of a reduced set of $n = 6$ redundant coordinates.

Next, the above formulation of the rotary crane is linked to the original one (see also [2, 31]) relying on the reduced set of crane coordinates (Fig. 5.11)

$$\overline{\boldsymbol{p}} = \begin{bmatrix} \varphi & s & l \end{bmatrix}^T \tag{5.134}$$

Here, the angle φ measures the rotation of the girder bridge about the Z-axis relative to the X-axis, the displacement s specifies the position of the trolley on the girder bridge, and the variable l denotes the length of the hoisting cable connecting the load with the winch contained in the trolley. In contrast to the previous crane model in Fig. 5.10, the winch contained in the trolley is now assumed to be actuated (torque M_w). The previous crane coordinates $\boldsymbol{p}$ in (5.120) can now be expressed in terms of the reduced set of crane coordinates (5.134) and this gives rise to the mapping $\boldsymbol{p} = \boldsymbol{\varphi}(\overline{\boldsymbol{p}})$ in (5.111). Accordingly,

$$\begin{bmatrix} x_2 \\ y_2 \\ x_0 \\ y_0 \\ L_1 \\ L_2 \\ L_0 \end{bmatrix} = \begin{bmatrix} -r\cos\varphi \\ -r\sin\varphi \\ s\cos\varphi \\ s\sin\varphi \\ s+\alpha r \\ s+r+l \\ s+r \end{bmatrix} \quad \text{and} \quad D\boldsymbol{\varphi}(\boldsymbol{p}) = \begin{bmatrix} r\sin\varphi & 0 & 0 \\ -r\cos\varphi & 0 & 0 \\ -s\sin\varphi & \cos\varphi & 0 \\ s\cos\varphi & \sin\varphi & 0 \\ 0 & 1 & 0 \\ 0 & 1 & 1 \\ 0 & 1 & 0 \end{bmatrix} \tag{5.135}$$

Furthermore, the quantities in (5.114) can now be calculated in a straightforward way and it leads to

$$\overline{\boldsymbol{M}}_1(\overline{\boldsymbol{p}}) = \begin{bmatrix} Mr^2 + m_0 s^2 & 0 & 0 \\ 0 & m_0 + \frac{J_1}{r_1^2} + \frac{J_2}{r_2^2} & \frac{J_2}{r_2^2} \\ 0 & \frac{J_2}{r_2^2} & \frac{J_2}{r_2^2} + \frac{J_w}{r_w^2} \end{bmatrix} \tag{5.136}$$

and

$$\overline{\boldsymbol{f}}_1(\overline{\boldsymbol{p}}, \dot{\overline{\boldsymbol{p}}}) = \begin{bmatrix} -2m_0 s\dot{s}\dot{\varphi} \\ m_0 s\dot{\varphi}^2 \\ 0 \end{bmatrix}, \qquad \overline{\boldsymbol{B}}_1^T(\overline{\boldsymbol{p}}) = \begin{bmatrix} 0 & 0 & 1 \\ \frac{1}{r_1} & \frac{1}{r_2} & 0 \\ 0 & \frac{1}{r_2} & 0 \end{bmatrix} \tag{5.137}$$

There remains one holonomic constraint (5.113b) which is given by

$$\overline{\Phi}(\overline{\boldsymbol{p}}, \boldsymbol{x}) = \frac{1}{2}((x - s\cos\varphi)^2 + (y - s\sin\varphi)^2 + z^2 - l^2) = 0 \tag{5.138}$$

Finally, note that the control inputs

$$\overline{\boldsymbol{u}} = \begin{bmatrix} M_b \\ F_t \\ \frac{M_w}{r_w} \end{bmatrix} \tag{5.139}$$

conjugated to the reduced crane coordinates (5.134) can be obtained from $\overline{\boldsymbol{u}} = \overline{\boldsymbol{B}}_1^T(\overline{\boldsymbol{p}})\boldsymbol{u}$. In particular,

$$F_t = \frac{M_1}{r_1} + \frac{M_2}{r_2} \tag{5.140a}$$

$$M_w = \frac{r_w}{r_2} M_2 \tag{5.140b}$$

That is, the two winch torques M_1 and M_2 of the original model are linked to the force F_t acting on the trolley and the winch torque M_w (cf. Fig. 5.11).

Numerical results

In the numerical simulation, the same data are used as in [2, 28]. In particular, the inertia parameters are given by $m = 100\,\text{kg}$, $m_0 = 10\,\text{kg}$, $J_b = 480\,\text{kg}\cdot\text{m}^2$, and $M = \frac{J_b}{r^2}$ with $r = 4\,\text{m}$. Concerning the moment of inertia corresponding to the winches, $J_w = 0.1\,\text{kg}\cdot\text{m}^2$ and $J_1 = J_2 = 0$ are chosen. Moreover, $r_w = 0.1\,\text{m}$.

The servo constraints are used to prescribe a rest-to-rest maneuver of the load specified by

$$\gamma(t) = \gamma_0 + (\gamma_f - \gamma_0)c(t) \tag{5.141}$$

with the initial position

$$\gamma_0 = \begin{bmatrix} 5\,\text{m} \\ 0 \\ -5\,\text{m} \end{bmatrix} \quad \text{at} \quad t_0 = 0 \tag{5.142}$$

and the final position

$$\gamma_f = \begin{bmatrix} -2\,\text{m} \\ 2\,\text{m} \\ -2\,\text{m} \end{bmatrix} \quad \text{at} \quad t_f = 20\,\text{s} \tag{5.143}$$

The reference function $c(t)$ is composed of three phases,

$$c(t) = \begin{cases} c_I(t) & \text{for } 0 \leq t < 5\,\text{s} \\ c_{II}(t) & \text{for } 5\,\text{s} \leq t < 15\,\text{s} \\ c_{III}(t) & \text{for } 15\,\text{s} \leq t \leq 20\,\text{s} \end{cases}$$

with each phase

$$\begin{aligned} c_I(t) &= \frac{1}{\tau - \tau_0}\left(-\frac{5t^8}{2\tau_0{}^7} + \frac{10t^7}{\tau_0{}^6} - \frac{14t^6}{2\tau_0{}^5} + \frac{7t^5}{2\tau_0{}^4}\right) \\ c_{II}(t) &= \frac{1}{\tau - \tau_0}\left(t - \frac{\tau_0}{2}\right) \\ c_{III}(t) &= 1 + \frac{1}{\tau - \tau_0}\left(-\frac{5(\tau - t)^8}{2\tau_0{}^7} + \frac{10(\tau - t)^7}{\tau_0{}^6} - \frac{14(\tau - t)^6}{2\tau_0{}^5} + \frac{7(\tau - t)^5}{2\tau_0{}^4}\right) \end{aligned}$$

Using the reduced crane coordinates, the initial configuration of the rotary crane at $t_0 = 0$ is defined by $\overline{\boldsymbol{p}}_0 = \begin{bmatrix} 0 & 5\,\text{m} & 5\,\text{m} \end{bmatrix}^T$, while the initial load coordinates are given by $\boldsymbol{x}_0 = \begin{bmatrix} 5\,\text{m} & 0 & -5\,\text{m} \end{bmatrix}^T$. The motion of the crane is starting at rest such that $\dot{\boldsymbol{\mu}}_0 = \boldsymbol{0}$.

In Fig. 5.12 and 5.13, the numerical solution (NUM) is compared to the analytical reference solution (REF) obtained in [2]. It can be observed that the numerical solution converges to the reference solution when the time step size is reduced. In addition to that, Fig. 5.14 displays the numerical solution for the extended crane coordinates $\boldsymbol{p}$. The two alternative formulations in terms of redundant coordinates ($\boldsymbol{p}$ and $\overline{\boldsymbol{p}}$, respectively) yield practically indistinguishable results. The simulated motion of the rotary crane in terms of extended crane coordinates $\boldsymbol{p}$ is illustrated in Fig. 5.15 with some snapshots at consecutive points in time. Similarly, snapshots obtained with the formulation in terms of the reduced crane coordinates $\overline{\boldsymbol{p}}$ are shown in Fig. 5.16.

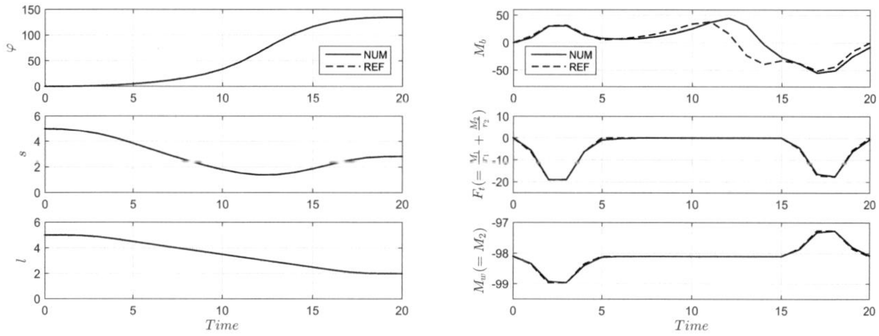

Figure 5.12: Rotary crane: Comparison between the numerical results (NUM) obtained with $\Delta t = 1\,\mathrm{s}$ and the reference solution (REF) for the reduced crane coordinates $\overline{\boldsymbol{p}}(t)$.

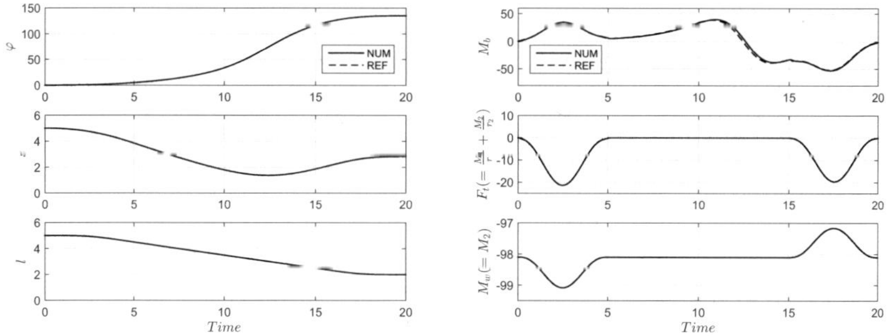

Figure 5.13: Rotary crane: Comparison between the numerical results (NUM) obtained with $\Delta t = 0.1\,\mathrm{s}$ and the reference solution (REF) for the reduced crane coordinates $\overline{\boldsymbol{p}}(t)$.

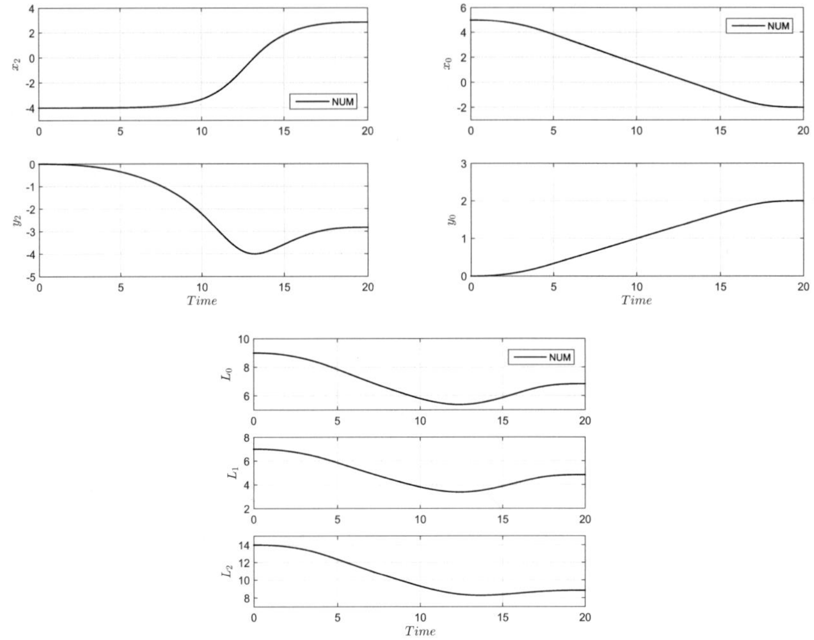

Figure 5.14: Rotary crane: Numerical results (NUM) for the extended crane coordinates $\boldsymbol{p}(t)$ obtained with $\Delta t = 0.1\,\mathrm{s}$.

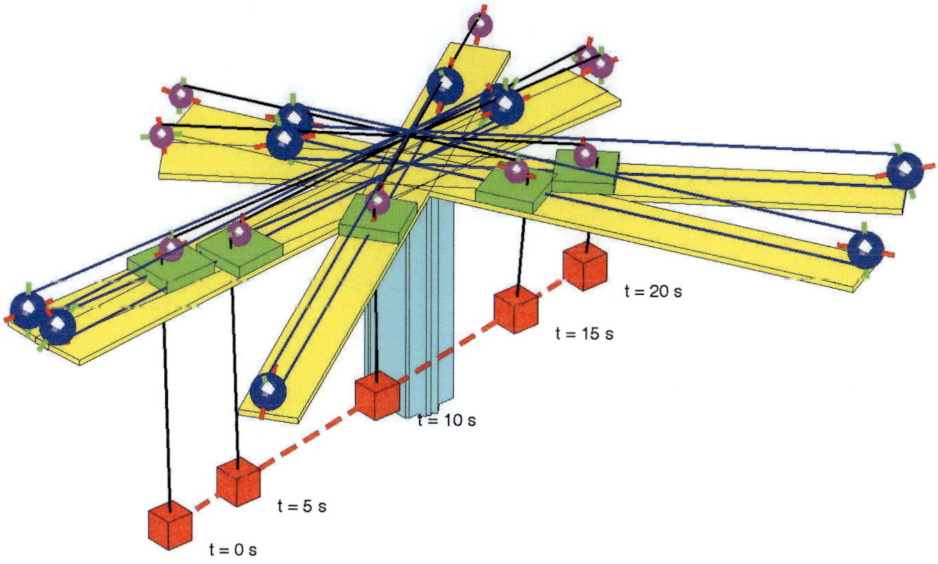

Figure 5.15: Rotary crane (formulation in terms of the extended crane coordinates $\boldsymbol{p}$): Snapshots at specific points in time.

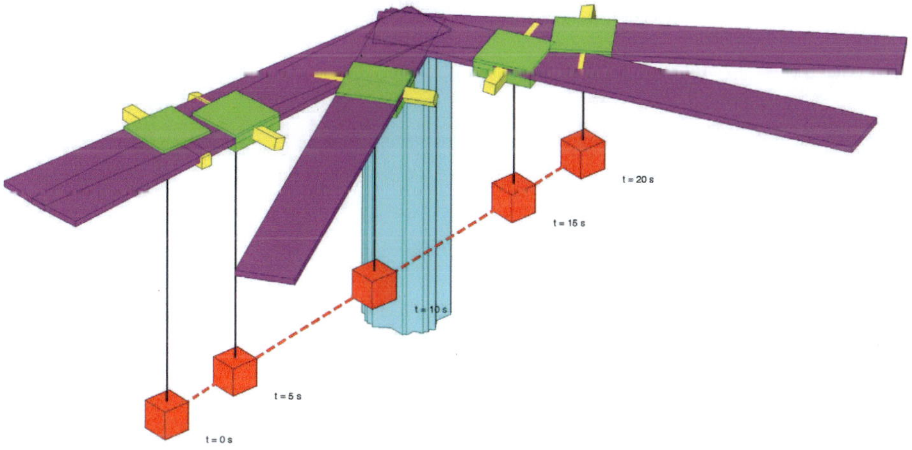

Figure 5.16: Rotary crane (formulation in terms of the reduced crane coordinates $\overline{\boldsymbol{p}}$): Snapshots at specific points in time.

6 Index reduction by minimal extension for advanced examples

In this chapter the effective application of index reduction by minimal extension will be demonstrated by several advanced examples of underactuated systems, for which the formulation in terms of redundant coordinates is employed within the inverse dynamics simulation. The numerical examples are the planar US Navy crane, the three-dimensional US Navy crane and a cable suspension manipulator.

6.1 Planar US Navy crane

The example of US Navy crane in the plane was originally treated as a crane control problem in [65]. It is a simplified version of the crane used by the US Navy.

6.1.1 Planar US Navy crane with neglected pulley mass

The case, in which the mass of the mobile pulley at point B (see Fig. 6.1) is neglected ($m_0 = 0$), is first considered.

As illustrated in Fig. 6.1, the crane consists of a pole and a system of two cables actuated by two winches and linked by a mobile pulley. The pole is assumed to make a fixed angle α with respect to the vertical, and is equipped with two winches, one located at the origin P, and the other located at point A, at a fixed distance l from point P.

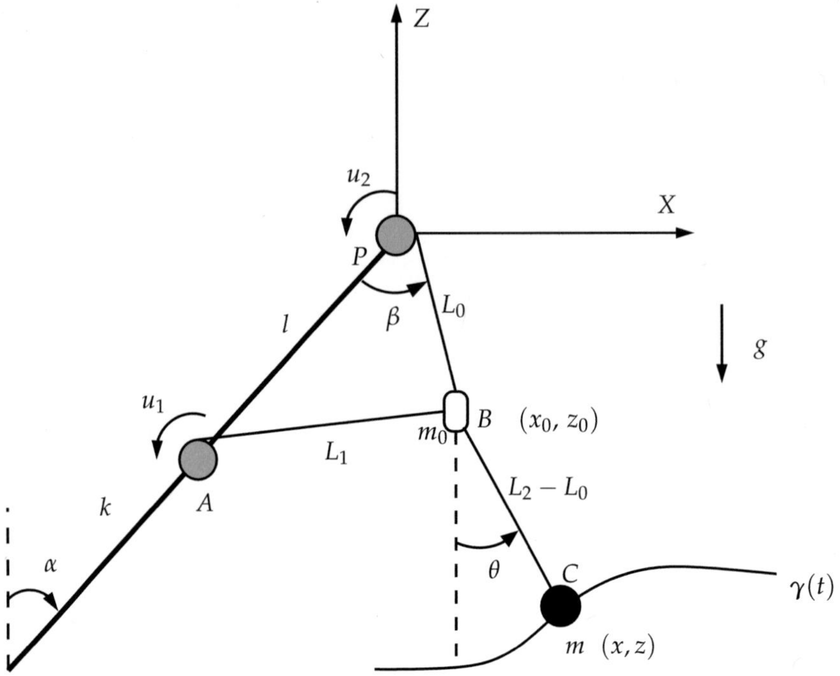

Figure 6.1: The planar US Navy crane model with neglected pully mass ($m_0 = 0$) in terms of $n = 7$ redundant coordinates.

The first cable of variable length L_2, whose upper part of variable length L_0 makes an angle β with the pole and whose lower part of variable length $L_2 - L_0$ makes an angle θ with the vertical, starts from the winch at point P (radius r_2, moment of inertia J_2, actuating torque u_2), passes through the mobile pulley located at point B, and ends up on the load (mass m) located at point C. The second cable of variable length L_1 relates the winch at point A (radius r_1, moment of inertia J_1, actuating torque u_1) to the pulley at point B. All the cables are assumed to be massless and unstretchable.

Redundant coordinates formulation

The very general framework in [58] can be used to model the cranes. Accordingly, $n = 7$ redundant coordinates subjected to $m = 3$ holonomic constraints are used. The enlarged set of redundant crane coordinates (see Fig. 6.1) is given by

$$p = \begin{bmatrix} L_1 & L_2 & L_0 & x_0 & z_0 \end{bmatrix}^T \tag{6.1}$$

and

$$x = \begin{bmatrix} x & z \end{bmatrix}^T \tag{6.2}$$

Similarly, the coordinates of the load (mass m) are described in a Cartesian inertial frame such that the servo constraints can be expressed in a simple manner. The holonomic constraints $\boldsymbol{h}(\boldsymbol{p}) = \boldsymbol{0}$ are given by

$$\boldsymbol{h}(\boldsymbol{p}) = \begin{bmatrix} \frac{1}{2}\left((x_0 + l\sin\alpha)^2 + (z_0 + l\cos\alpha)^2 - L_1^2\right) \\ \frac{1}{2}\left(x_0^2 + z_0^2 - L_0^2\right) \end{bmatrix} \tag{6.3}$$

Accordingly, $m_1 = 2$. The first constraint links the coordinate L_1 to the position of pulley and the position of winch at point A, and the second constraint links the coordinate L_0 to the position of pulley and the position of winch at point P. Moreover, the holonomic constraint $\boldsymbol{\Phi}(\boldsymbol{p}, \boldsymbol{x}) = \boldsymbol{0}$ is specified by

$$\Phi(\boldsymbol{p}, \boldsymbol{x}) = \frac{1}{2}((x - x_0)^2 + (z - z_0)^2 - (L_2 - L_0)^2) \tag{6.4}$$

and relates the load coordinates to the crane (or robot) coordinates. Accordingly, $m_2 = 1$. The total kinetic energy of the crane system under consideration can be written in the form

$$T = \frac{1}{2}\dot{\boldsymbol{p}} \cdot \boldsymbol{M}_1 \dot{\boldsymbol{p}} + \frac{1}{2}\dot{\boldsymbol{x}} \cdot \boldsymbol{M}_2 \dot{\boldsymbol{x}} \tag{6.5}$$

in which the mass matrices corresponding to the crane coordinates and the load coordinates are given by

$$\boldsymbol{M}_1 = \begin{bmatrix} \frac{J_1}{r_1^2} & 0 & 0 & 0 & 0 \\ 0 & \frac{J_2}{r_2^2} & 0 & 0 & 0 \\ 0 & 0 & 0 & 0 & 0 \\ 0 & 0 & 0 & 0 & 0 \\ 0 & 0 & 0 & 0 & 0 \end{bmatrix}, \qquad \boldsymbol{M}_2 = \begin{bmatrix} m & 0 \\ 0 & m \end{bmatrix} \tag{6.6}$$

Further quantities needed in Equations (5.102a)–(5.102d) are given by

$$\boldsymbol{B}_1^T = \begin{bmatrix} \frac{1}{r_1} & 0 \\ 0 & \frac{1}{r_2} \\ 0 & 0 \\ 0 & 0 \\ 0 & 0 \end{bmatrix}, \quad \boldsymbol{f}_1 = \begin{bmatrix} 0 \\ 0 \\ 0 \\ 0 \\ 0 \end{bmatrix}, \quad \boldsymbol{f}_2 = \begin{bmatrix} 0 \\ -mg \end{bmatrix}, \quad \boldsymbol{u} = \begin{bmatrix} u_1 \\ u_2 \end{bmatrix} \tag{6.7}$$

and by

$$
\boldsymbol{H}_1^T = \begin{bmatrix} -L_1 & 0 \\ 0 & 0 \\ 0 & -L_0 \\ x_0 + l\sin\alpha & x_0 \\ z_0 + l\cos\alpha & z_0 \end{bmatrix}, \qquad \boldsymbol{G}_1^T = \begin{bmatrix} 0 \\ L_0 - L_2 \\ L_2 - L_0 \\ x_0 - x \\ z_0 - z \end{bmatrix}, \qquad \boldsymbol{G}_2^T = \begin{bmatrix} x - x_0 \\ z - z_0 \end{bmatrix} \tag{6.8}
$$

To summarize, the crane system has $f = 4$ degrees of freedom and $a = 2$ control inputs given by the two winch torques u_1 and u_2. Therefore, the US Navy crane at hand is a typical underactuated mechanical system. The equations of motion are given in detail in Appendix A.2.

Analytical solution based on differential flatness

The US Navy crane can be classified as differentially flat system. Then by proceeding along the lines of Lévine et al. [65] one obtains the analytical reference solution based on differential flatness. This means that all the system variables can be expressed as functions of the load coordinates (also flat outputs) and a finite number of their derivatives.

Inverse dynamics simulation

The numerical simulation is performed with the following parameters: $m = 100\,\text{kg}$, $J_1 = J_2 = 0.1\,\text{kg}\cdot\text{m}^2$, $r_1 = r_2 = 0.1\,\text{m}$, $\alpha = \frac{\pi}{3}$, and $l = 10\,\text{m}$. The partially specified motion of the load is rest-to-rest and the same functions are used to generate the prescribed trajectory as in the example of the planar overhead crane. The initial position is specified by

$$
\gamma_0 = \begin{bmatrix} 0 & -15\,\text{m} \end{bmatrix}^T \quad \text{at} \quad t = 0 \tag{6.9}
$$

and the final position is given by

$$
\gamma_f = \begin{bmatrix} -5\,\text{m} & -12\,\text{m} \end{bmatrix}^T \quad \text{at} \quad t = 3\,\text{s} \tag{6.10}
$$

The initial configuration of the crane is defined by

$$
\boldsymbol{p} = \begin{bmatrix} 5\sqrt{3}\,\text{m} & 15\,\text{m} & 5\,\text{m} & 0 & -5\,\text{m} \end{bmatrix}^T \tag{6.11}
$$

at $t_0 = 0$, and the initial load coordinates are given by

$$x = \begin{bmatrix} 0 & -15\,\mathrm{m} \end{bmatrix}^T \tag{6.12}$$

As shown in Fig. 6.2, the numerical results of the coordinates are identical to the analytical reference solution even for the coarse time step size $\Delta t = 0.1\,\mathrm{s}$. This implies that the numerical results based on redundant coordinates are independent from the selected time step size due to the property of differential flatness. Fig. 6.3 shows that the numerical solution of the control inputs converges to the reference solution when the time step size is reduced, and Fig. 6.4 displays the numerical solution of the Lagrange multipliers for the time step size $\Delta t = 0.01\,\mathrm{s}$. The simulated motion of the crane in terms of redundant coordinates is presented in Fig. 6.5 with some snapshots at consecutive points in time. Note that the cable of variable length L_1 is horizontal in the initial configuration since the pulley mass is neglected, otherwise a third cable is needed to suspend the mobile pulley from its right hand side (see Fig. 6.6).

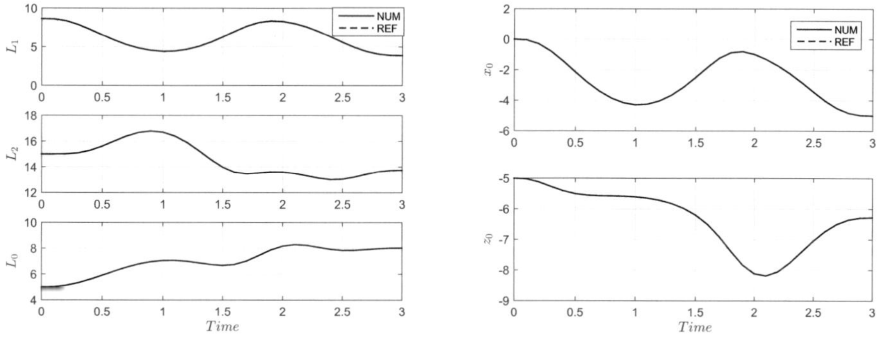

Figure 6.2: Planar US Navy crane with neglected pulley mass: Comparison between numerical results (NUM) obtained with $\Delta t = 0.1\,\mathrm{s}$ and the analytical reference solution (REF).

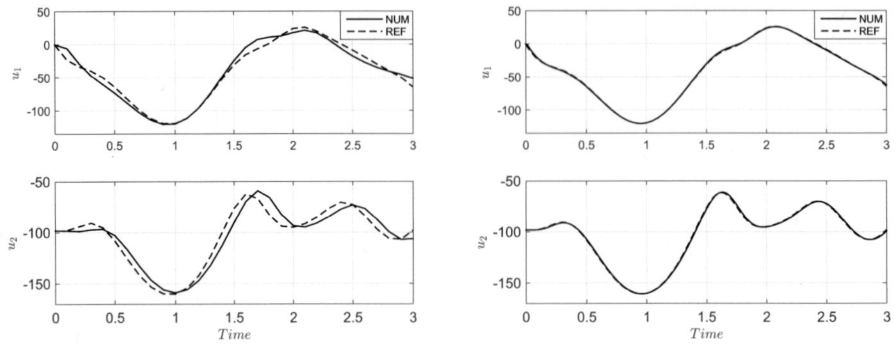

Figure 6.3: Planar US Navy crane with neglected pulley mass: Comparison between numerical results (NUM) with $\Delta t = 0.1\,\mathrm{s}$ (left Fig.) and $\Delta t = 0.01\,\mathrm{s}$ (right Fig.) and the analytical reference solution (REF).

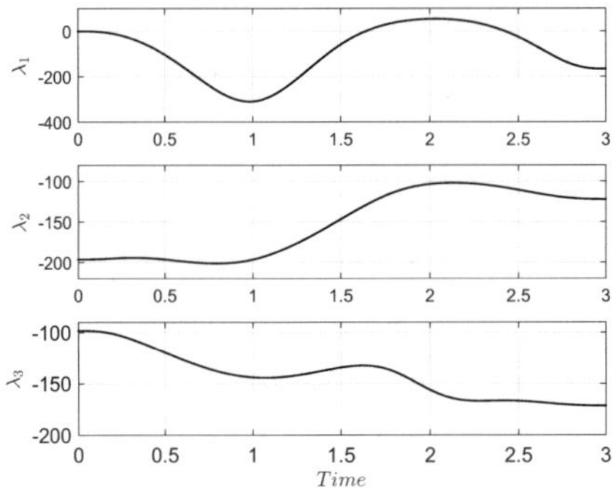

Figure 6.4: Planar US Navy crane with neglected pulley mass: Numerical results (NUM) of the Lagrange multipliers obtained with $\Delta t = 0.01\,\mathrm{s}$.

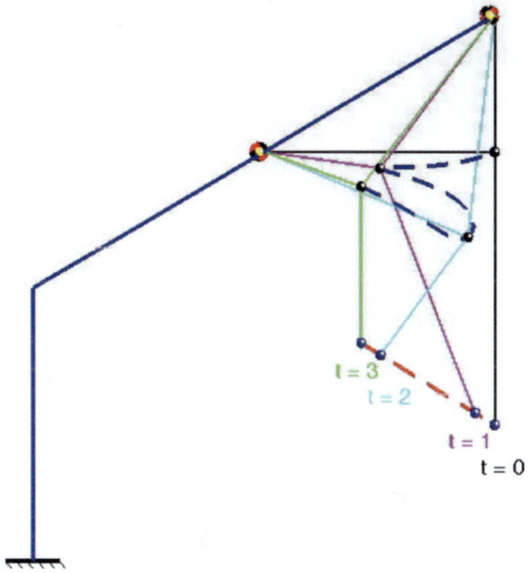

Figure 6.5: Planar US Navy crane with neglected pulley mass: Snapshots of the load mass and the pulley at specific points in time. Besides the trajectory of pulley and the prescribed trajectory of load mass is shown.

6.1.2 Planar US Navy crane with nonzero pulley mass

Next, we consider the case in which the mass of the mobile pulley at point B (see Fig. 6.6) is nonzero ($m_0 > 0$).

As illustrated in Fig. 6.6, the crane consists of a pole and a system of three cables actuated by three winches and linked by a mobile pulley. The pole is assumed to make a fixed angle α with respect to the vertical, and is equipped with three winches, one located at the origin S, the second located at point P, at a fixed distance s from point S, and the third located at point A, at a fixed distance l from point P.

The vertical cable of variable length L_2, whose upper part of variable length L_0 makes an angle β with the pole and whose lower part of variable length $L_2 - L_0$ makes an angle θ with the vertical, starts from the winch at point P (radius r_2, moment of inertia J_2, actuating torque u_2), passes through the mobile pulley located at point B, and ends up on the load (mass m) located at point C.

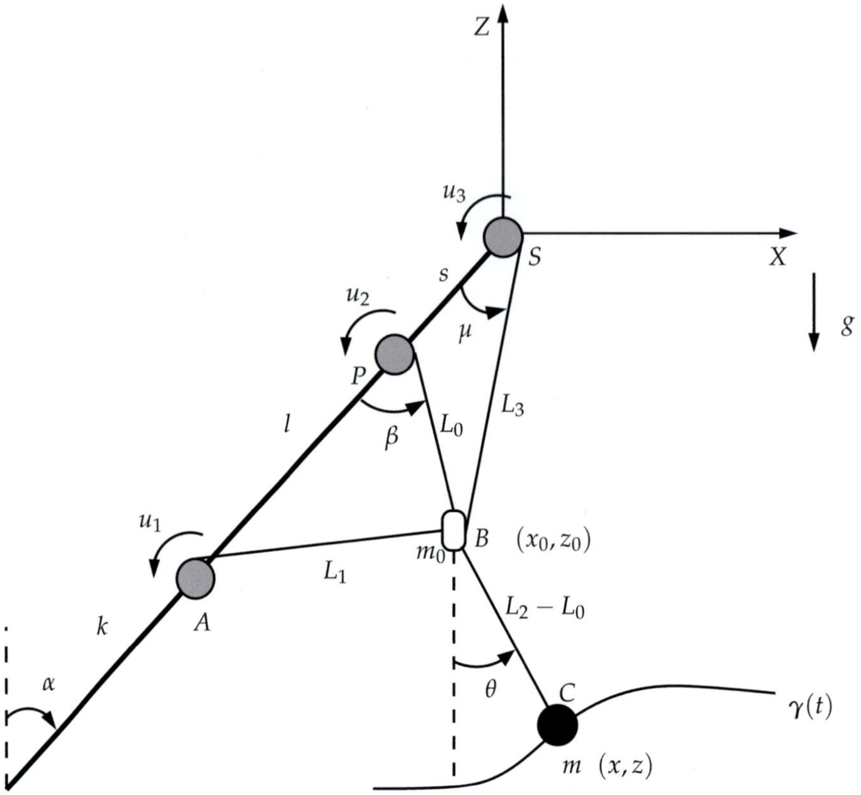

Figure 6.6: The planar US Navy crane model with nonzero pully mass ($m_0 > 0$) in terms of $n = 8$ redundant coordinates.

The second cable of variable length L_1 relates the winch at point A (radius r_1, moment of inertia J_1, actuating torque u_1) to the pulley at point B. The suspension cable for the mobile pulley of variable length L_3 starts from the winch at point S (radius r_3, moment of inertia J_3, actuating torque u_3), makes an angle μ with the pole, and ends at the free pulley at point B. All the cables are assumed to be massless and unstretchable. Note that the number of control inputs is now increased by one, the new input u_3 is required to hoist the suspension cable of the mobile pulley. Nevertheless, the flatness property is still conserved, which has been proved in the dissertation of Kiss [57].

Redundant coordinates formulation

According to the general framework in [58], the mathematical model of the planar US Navy crane with nonzero pulley mass can be formulated in terms of $n = 8$ redundant coordinates, which are subject to $m = 4$ holonomic constraints. The enlarged set of redundant crane coordinates, as depicted in Fig. 6.6, is expressed by

$$p = \begin{bmatrix} L_1 & L_2 & L_3 & L_0 & x_0 \end{bmatrix}^T \tag{6.13}$$

and by

$$x = \begin{bmatrix} x & z & z_0 \end{bmatrix}^T \tag{6.14}$$

Note that the coordinate z_0 of the mobile pulley is selected as the third flat output, since there are three control inputs for the crane system at hand. In addition, other possible choices for the third flat output could be the cable length variable $L_2 - L_0$ or the coordinate x_0 of the mobile pulley.

The holonomic constraints $\boldsymbol{h}(\boldsymbol{p}) = \boldsymbol{0}$ vanish, since the coordinate z_0 is present in each constraint equation. Accordingly, $m_1 = 0$. Moreover, the holonomic constraints $\boldsymbol{\Phi}(\boldsymbol{p}, \boldsymbol{x}) = \boldsymbol{0}$ are given by

$$\boldsymbol{\Phi}(\boldsymbol{p}, \boldsymbol{x}) = \begin{bmatrix} \frac{1}{2}((x_0 + (l+s)\sin\alpha)^2 + (z_0 + (l+s)\cos\alpha)^2 - L_1^2) \\ \frac{1}{2}((x_0 + s\,\sin\alpha)^2 + (z_0 + s\,\cos\alpha)^2 - L_0^2) \\ \frac{1}{2}(x_0^2 + z_0^2 - L_3^2) \\ \frac{1}{2}((x - x_0)^2 + (z - z_0)^2 - (L_2 - L_0)^2) \end{bmatrix} \tag{6.15}$$

The first constraint links the coordinate L_1 to the position of pulley and the position of winch at point A, the second constraint links the coordinate L_0 to the position of pulley and the position of winch at point P, the third constraint links the coordinate L_3 to the position of pulley and the position of winch at point S, and the fourth constraint connects the load coordinates with the position of pulley.

Accordingly, $m_2 = 4$. The total kinetic energy of the crane system assumes the form

$$T = \frac{1}{2}\dot{\boldsymbol{p}} \cdot \boldsymbol{M}_1 \dot{\boldsymbol{p}} + \frac{1}{2}\dot{\boldsymbol{x}} \cdot \boldsymbol{M}_2 \dot{\boldsymbol{x}} \tag{6.16}$$

in which the mass matrices corresponding to the crane coordinates and the extended load coordinates in Equation (6.14) are given by

$$M_1 = \begin{bmatrix} \frac{J_1}{r_1^2} & 0 & 0 & 0 & 0 \\ 0 & \frac{J_2}{r_2^2} & 0 & 0 & 0 \\ 0 & 0 & \frac{J_3}{r_3^2} & 0 & 0 \\ 0 & 0 & 0 & 0 & 0 \\ 0 & 0 & 0 & 0 & m_0 \end{bmatrix}, \qquad M_2 = \begin{bmatrix} m & 0 & 0 \\ 0 & m & 0 \\ 0 & 0 & m_0 \end{bmatrix} \tag{6.17}$$

Further quantities needed in Equations (5.102a)–(5.102d) are given by

$$B_1^T = \begin{bmatrix} \frac{1}{r_1} & 0 & 0 \\ 0 & \frac{1}{r_2} & 0 \\ 0 & 0 & \frac{1}{r_3} \\ 0 & 0 & 0 \\ 0 & 0 & 0 \end{bmatrix}, \quad f_1 = \begin{bmatrix} 0 \\ 0 \\ 0 \\ 0 \\ 0 \end{bmatrix}, \quad f_2 = \begin{bmatrix} 0 \\ -mg \\ -m_0 g \end{bmatrix}, \quad u = \begin{bmatrix} u_1 \\ u_2 \\ u_3 \end{bmatrix}. \tag{6.18}$$

and by

$$G_1^T = \begin{bmatrix} -L_1 & 0 & 0 & 0 \\ 0 & 0 & 0 & L_0 - L_2 \\ 0 & 0 & -L_3 & 0 \\ 0 & -L_0 & 0 & L_2 - L_0 \\ x_0 + (l+s)\sin\alpha & x_0 + s\sin\alpha & x_0 & x_0 - x \end{bmatrix} \tag{6.19}$$

$$G_2^T = \begin{bmatrix} 0 & 0 & 0 & x - x_0 \\ 0 & 0 & 0 & z - z_0 \\ z_0 + (l+s)\cos\alpha & z_0 + s\cos\alpha & z_0 & z_0 - z \end{bmatrix} \tag{6.20}$$

Analytical solution based on differential flatness

The US Navy crane with nonzero pulley mass has $f = 4$ degrees of freedom and $a = 3$ control inputs. As has been mentioned before, this system can also be classified as a differentially flat system. Accordingly, all the system variables can be expressed as functions of the flat output x and its derivatives up to a certain order. The derivation of the analytical solution, which has been applied in the case of $m_0 = 0$, can also be used here to provide the reference solution for the case of $m_0 > 0$. An alternative approach to the method given in Lévine et al. [65] is introduced to get the flatness-based solution. In this connection, the equations of motion for the crane model at hand can be given as follows:

$$\frac{J_1}{r_1^2}\ddot{L}_1 = -\lambda_1 L_1 + \frac{u_1}{r_1} \tag{6.21a}$$

$$\frac{J_2}{r_2^2}\ddot{L}_2 = \lambda_4(L_0 - L_2) + \frac{u_2}{r_2} \tag{6.21b}$$

$$\frac{J_3}{r_3^2}\ddot{L}_3 = -\lambda_3 L_3 + \frac{u_3}{r_3} \tag{6.21c}$$

$$0 = -\lambda_2 L_0 + \lambda_4(L_2 - L_0) \tag{6.21d}$$

$$m_0\ddot{x}_0 = \lambda_1(x_0 + (l+s)\sin\alpha) + \lambda_2(x_0 + s\sin\alpha) + \lambda_3 x_0 + \lambda_4(x_0 - x) \tag{6.21e}$$

$$m_0\ddot{z}_0 = \lambda_1(z_0 + (l+s)\cos\alpha) + \lambda_2(z_0 + s\cos\alpha) + \lambda_3 z_0 + \lambda_4(z_0 - z) - m_0 g \tag{6.21f}$$

$$m\ddot{x} = \lambda_4(x - x_0) \tag{6.21g}$$

$$m\ddot{z} = \lambda_4(z - z_0) - mg \tag{6.21h}$$

$$0 = \frac{1}{2}((x_0 + (l+s)\sin\alpha)^2 + (z_0 + (l+s)\cos\alpha)^2 - L_1^2) \tag{6.21i}$$

$$0 = \frac{1}{2}((x_0 + s\,\sin\alpha)^2 + (z_0 + s\,\cos\alpha)^2 - L_0^2) \tag{6.21j}$$

$$0 = \frac{1}{2}(x_0^2 + z_0^2 - L_3^2) \tag{6.21k}$$

$$0 = \frac{1}{2}((x - x_0)^2 + (z - z_0)^2 - (L_2 - L_0)^2) \tag{6.21l}$$

$$x = \gamma_1(t) \tag{6.21m}$$

$$z = \gamma_2(t) \tag{6.21n}$$

$$z_0 = \gamma_3(t) \tag{6.21o}$$

Due to the differential flatness property, the analytical reference solution can be derived through purely algebraic manipulations from the above equations.

At first, λ_4 and x_0 can be obtained from the equations (6.21h) and (6.21g) as a funtion of $\boldsymbol{x}$ and $\ddot{\boldsymbol{x}}$. Then the equations (6.21i), (6.21j), (6.21k), (6.21l) and (6.21d) are used to express the variables L_1, L_0, L_3, L_2 and λ_2 as function of $\boldsymbol{x}$ and $\ddot{\boldsymbol{x}}$. Next λ_1 and λ_3 can be expressed as function of $\boldsymbol{x}$ and $\ddot{\boldsymbol{x}}$ from the equations (6.21e) and (6.21f). At last the equations (6.21a), (6.21b) and (6.21c) are used to express u_1, u_2, and u_3 as functions of $\boldsymbol{x}$, $\dot{\boldsymbol{x}}$, $\ddot{\boldsymbol{x}}$, $\boldsymbol{x}^{(3)}$ and $\boldsymbol{x}^{(4)}$.

Obviously the fourth order derivative of the flat output $\boldsymbol{x}$ is present in the expression of the flatness-based solution. Thus it can be concluded that the differential index of the DAEs (A.3) is five.

Inverse dynamics simulation

For the numerical simulation, the following parameters are given: $m = 100\,\mathrm{kg}$, $m_0 = 150\,\mathrm{kg}$, $J_1 = J_2 = J_3 = 0.1\,\mathrm{kg}\cdot\mathrm{m}^2$, $r = 0.1\,\mathrm{m}$, $\alpha = \frac{\pi}{3}$, $s = 5\,\mathrm{m}$, $l = 10\,\mathrm{m}$ and $k = 10\,\mathrm{m}$. The prescribed trajectory of the load can be obtained in the same way as in the case of neglected pulley mass. The initial position is given by

$$\gamma_0 = \begin{bmatrix} -\frac{5}{2}\sqrt{3}\,\mathrm{m} \\ -22.5\,\mathrm{m} \\ -12.5\,\mathrm{m} \end{bmatrix} \quad \text{at} \quad t = 0 \tag{6.22}$$

and the final position is given by

$$\gamma_f = \begin{bmatrix} -1\,\mathrm{m} \\ -17.5\,\mathrm{m} \\ -10.5\,\mathrm{m} \end{bmatrix} \quad \text{at} \quad t = 3\,\mathrm{s} \tag{6.23}$$

The initial configuration of the crane model is specified by

$$p = \begin{bmatrix} 10\,\mathrm{m} & 20\,\mathrm{m} & 13.23\,\mathrm{m} & 10\,\mathrm{m} & -\frac{5}{2}\sqrt{3}\,\mathrm{m} \end{bmatrix}^T \tag{6.24}$$

at $t_0 = 0$, and the initial load coordinates are given by

$$x = \begin{bmatrix} -\frac{5}{2}\sqrt{3}\,\mathrm{m} & -22.5\,\mathrm{m} & -12.5\,\mathrm{m} \end{bmatrix}^T \tag{6.25}$$

Fig. 6.7 shows that the numerical results of the coordinates coincide with the analytical reference solution for the coarse time step size $\Delta t = 0.1\,\mathrm{s}$. This implies that the numerical solution of redundant coordinates is independent from the chosen time step size due to the property of differential flatness. Fig. 6.8 shows that the numerical solution of the control inputs converges to the reference solution with the reduced time step size, and Fig. 6.9 displays the numerical solution of the Lagrange multipliers for the time step size $\Delta t = 0.01\,\mathrm{s}$. The simulated motion of the crane in terms of redundant coordinates is presented in Fig. 6.10 with some snapshots at consecutive points in time.

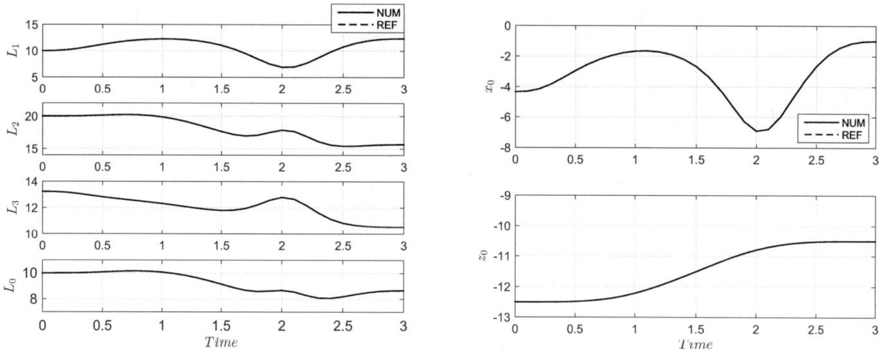

Figure 6.7: Planar US Navy crane with nonzero pulley mass: Comparison between the numerical results (NUM) obtained with $\Delta t = 0.1\,\mathrm{s}$ and the analytical reference solution (REF).

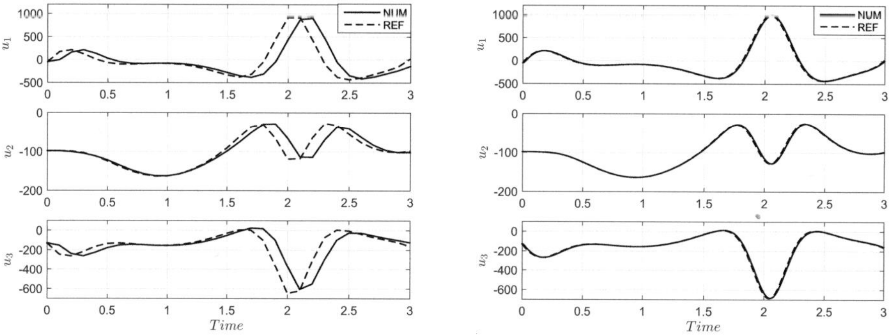

Figure 6.8: Planar US Navy crane with nonzero pulley mass: Comparison between the numerical results (NUM) obtained with $\Delta t = 0.1\,\mathrm{s}$ (left Fig.) and $\Delta t = 0.01\,\mathrm{s}$ (right Fig.) and the analytical reference solution (REF).

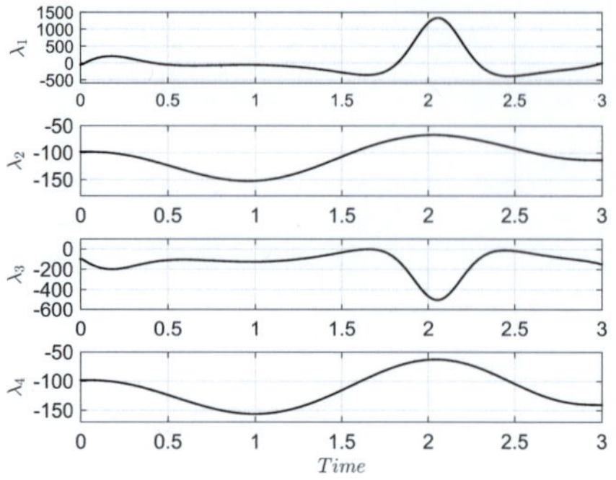

Figure 6.9: Planar US Navy crane with nonzero pulley mass: Numerical results (NUM) of the Lagrange multipliers obtained with $\Delta t = 0.01\,\mathrm{s}$.

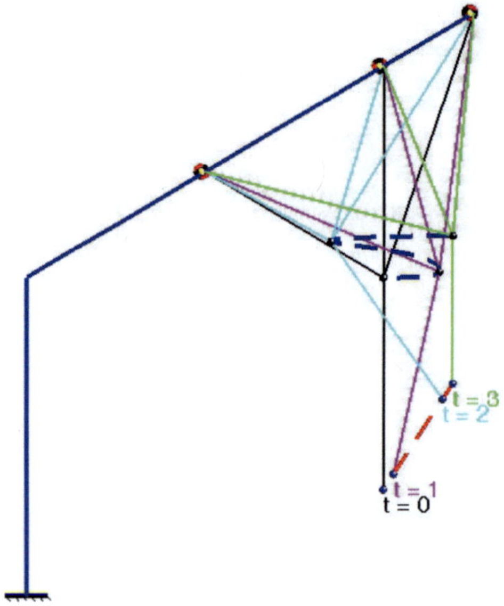

Figure 6.10: Planar US Navy crane with nonzero pulley mass: Snapshots of the load mass and the pulley at specific points in time. Besides the trajectory of pulley and the prescribed trajectory of load mass is shown.

6.2 Three-dimensional US Navy crane

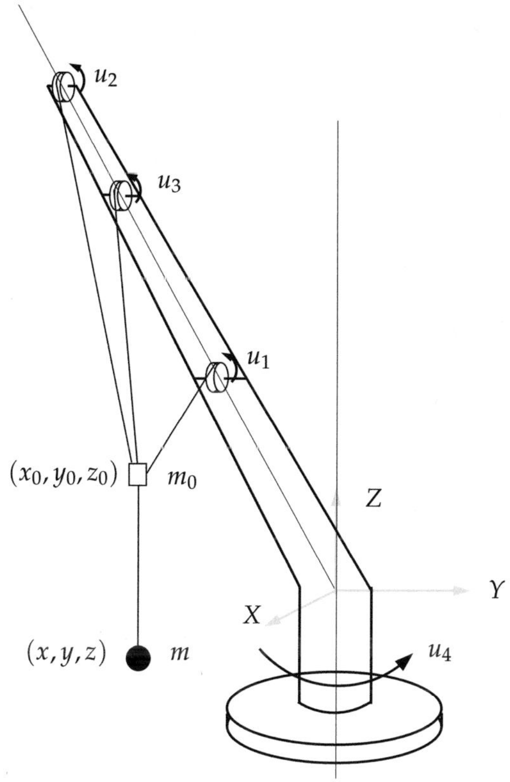

Figure 6.11: The three-dimensional US Navy crane model with nonzero pulley mass.

The planar US Navy crane can be extended to the case in three dimensions as depicted in Fig. 6.11. In three dimensions the whole mechanical system of the US Navy crane can rotate about the Z-axis of the inertial reference frame (see Fig. 6.11). Accordingly, additional variables, such as the actuating torque u_4, are needed to describe the mechanical state of the crane. In view of the mass of the free pulley, two cases are distinguished again: $m_0 = 0$ and $m_0 > 0$.

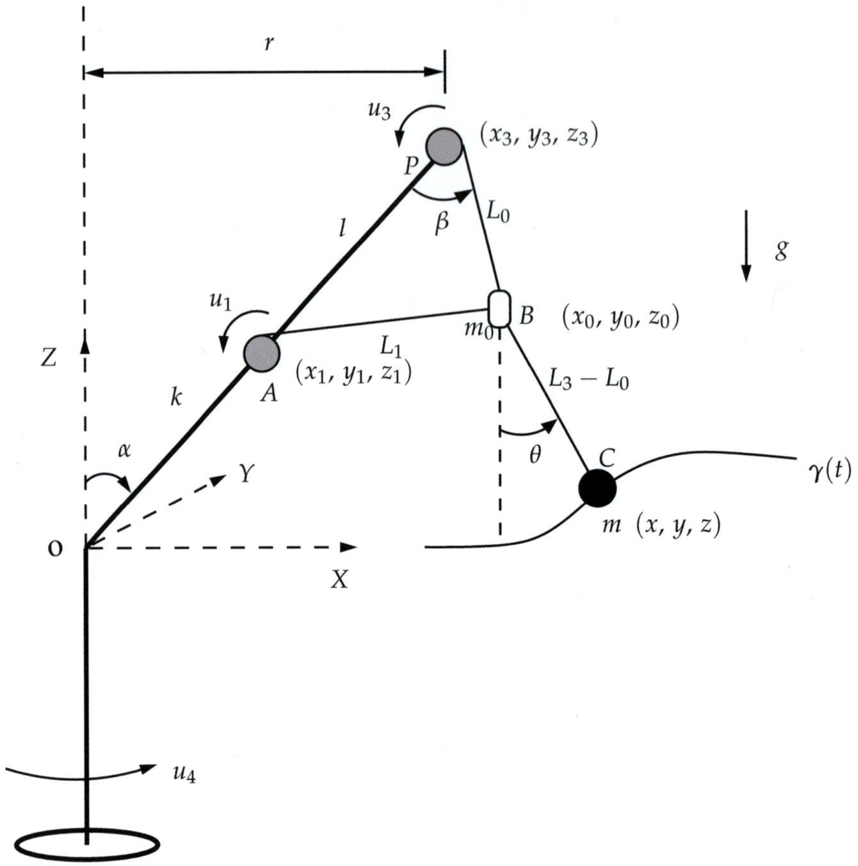

Figure 6.12: The three-dimensional US Navy crane model with neglected pulley mass ($m_0 = 0$) in terms of $n = 11$ redundant coordinates.

6.2.1 Three-dimensional US Navy crane with neglected pulley mass

At first, the case, in which the mass of the free pulley is neglected ($m_0 = 0$), is considered. Therefore, the suspension cable at the top position of the pole is slack and can be omitted.

As shown in Fig. 6.12, the crane consists of a pole and a system of two cables actuated by two winches and linked by a free pulley. The pole makes a fixed angle α with the vertical rotation axis (Z-axis) of the crane, and is equipped with two winches, one located at point P with Cartesian coordinates (x_3, y_3, z_3), and the other located at point A with Cartesian coordinates (x_1, y_1, z_1) , at a fixed distance l from point P (or at a fixed distance k from the origin O of the inertial reference frame).

The first cable of variable length L_3, whose upper part of variable length L_0 makes an angle β with the pole and whose lower part of variable length $L_3 - L_0$ makes an angle θ with the vertical, starts from the winch at point P (radius r_3, moment of inertia J_3, actuating torque u_3), passes through the free pulley located at point B, and ends up on the load (mass m) located at point C.

The second cable of variable length L_1 relates the winch at point A (radius r_1, moment of inertia J_1, actuating torque u_1) to the pulley at point B with Cartesian coordinates (x_0, y_0, z_0). All the cables are assumed to be massless and unstretchable.

It is obviously seen that the control input u_4 is newly required and makes the rotation of the whole crane system possible. The additional inertial parameter is the rotational inertia of the platform which is denoted by the mass M. Moreover, r denotes the perpendicular distance between the winch at point P and the rotation axis of the crane.

Redundant coordinates formulation

Proceeding along the lines in [58], $n = 11$ redundant coordinates subjected to $m = 4$ holonomic constraints are used to model the three-dimensional US Navy crane with neglected pulley mass. The enlarged set of redundant crane coordinates (see Fig. 6.12) is given by

$$\boldsymbol{p} = \begin{bmatrix} L_1 & L_3 & L_0 & x_0 & y_0 & z_0 & x_3 & y_3 \end{bmatrix}^T \tag{6.26}$$

and by

$$\boldsymbol{x} = \begin{bmatrix} x & y & z \end{bmatrix}^T \tag{6.27}$$

Note that the position of the winch at point A is fixed relative to the position of the winch at point P by a parameter $\beta_1 = \frac{1}{2}$. The holonomic constraints $\boldsymbol{h}(\boldsymbol{p}) = \boldsymbol{0}$ are given by

$$\boldsymbol{h}(\boldsymbol{p}) = \begin{bmatrix} \frac{1}{2}\left((x_0 - \beta_1 x_3)^2 + (y_0 - \beta_1 y_3)^2 + (z_0 - \beta_1 z_3)^2 - L_1^2\right) \\ \frac{1}{2}\left((x_0 - x_3)^2 + (y_0 - y_3)^2 + (z_0 - z_3)^2 - L_0^2\right) \\ \frac{1}{2}\left(x_3^2 + y_3^2 - r^2\right) \end{bmatrix} \tag{6.28}$$

Accordingly, $m_1 = 3$. The first constraint links the coordinate L_1 to the position of the pulley and the winch at point A, the second constraint links the coordinate L_0 to the position of the pulley and the winch at point P, and the third constraint links the parameter r to the position of the winch at point P. Furthermore, the holonomic constraint $\boldsymbol{\Phi}(\boldsymbol{p}, \boldsymbol{x}) = \boldsymbol{0}$ is specified by

$$\Phi(\boldsymbol{p}, \boldsymbol{x}) = \frac{1}{2}\left((x - x_0)^2 + (y - y_0)^2 + (z - z_0)^2 - (L_3 - L_0)^2\right) \tag{6.29}$$

and connects the load coordinates with the crane coordinates. Accordingly, $m_2 = 1$. The total kinetic energy of the crane system under consideration assumes the form

$$T = \frac{1}{2}\dot{\boldsymbol{p}} \cdot \boldsymbol{M}_1 \dot{\boldsymbol{p}} + \frac{1}{2}\dot{\boldsymbol{x}} \cdot \boldsymbol{M}_2 \dot{\boldsymbol{x}} \tag{6.30}$$

in which the mass matrices corresponding to the crane coordinates and the load coordinates are given by

$$\boldsymbol{M}_1 = \begin{bmatrix} \frac{J_1}{r_1^2} & & & & & & & \\ & \frac{J_3}{r_3^2} & & & & & & \\ & & 0 & & & & & \\ & & & 0 & & & & \\ & & & & 0 & & & \\ & & & & & 0 & & \\ & & & & & & M & \\ & & & & & & & M \end{bmatrix}, \quad \boldsymbol{M}_2 = \begin{bmatrix} m & & \\ & m & \\ & & m \end{bmatrix} \tag{6.31}$$

Further quantities needed in Equations (5.102a)–(5.102d) are given by

$$\boldsymbol{B}_1^T = \begin{bmatrix} \frac{1}{r_1} & 0 & 0 \\ 0 & \frac{1}{r_3} & 0 \\ 0 & 0 & 0 \\ 0 & 0 & 0 \\ 0 & 0 & 0 \\ 0 & 0 & 0 \\ 0 & 0 & -\frac{y_3}{r^2} \\ 0 & 0 & \frac{x_3}{r^2} \end{bmatrix}, \quad \boldsymbol{f}_1 = \begin{bmatrix} 0 \\ 0 \\ 0 \\ 0 \\ 0 \\ 0 \\ 0 \\ 0 \end{bmatrix}, \quad \boldsymbol{f}_2 = \begin{bmatrix} 0 \\ 0 \\ -mg \end{bmatrix}, \quad \boldsymbol{u} = \begin{bmatrix} u_1 \\ u_3 \\ u_4 \end{bmatrix} \tag{6.32}$$

and by

$$H_1^T = \begin{bmatrix} -L_1 & 0 & 0 \\ 0 & 0 & 0 \\ 0 & -L_0 & 0 \\ x_0 - \beta_1 x_3 & x_0 - x_3 & 0 \\ y_0 - \beta_1 y_3 & y_0 - y_3 & 0 \\ z_0 - \beta_1 z_3 & z_0 - z_3 & 0 \\ -\beta_1(x_0 - \beta_1 x_3) & x_3 - x_0 & x_3 \\ -\beta_1(y_0 - \beta_1 y_3) & y_3 - y_0 & y_3 \end{bmatrix}, \quad G_1^T = \begin{bmatrix} 0 \\ L_0 - L_3 \\ L_3 - L_0 \\ x_0 - x \\ y_0 - y \\ z_0 - z \\ 0 \\ 0 \end{bmatrix}, \quad G_2^T = \begin{bmatrix} x - x_0 \\ y - y_0 \\ z - z_0 \end{bmatrix} \tag{6.33}$$

In summary, the three-dimensional US Navy crane with neglected pulley mass has $f = 7$ degrees of freedom, $a = 3$ control inputs and can be classified as differentially flat system. Nevertheless, the flatness-based solution is much more complicated to be derived in the three-dimensional case. The equations of motion are given in detail in Appendix A.4.

Inverse dynamics simulation

The numerical experiment makes use of the following parameters: $m = 100\,\text{kg}$, $M = 6.4\,\text{kg}$, $J_1 = J_3 = 0.1\,\text{kg} \cdot \text{m}^2$, $r_1 = r_3 = 0.1\,\text{m}$, $\alpha = \frac{\pi}{3}$, and $l = k = 5\,\text{m}$. The prescribed trajectory can be calculated in the same way as in the three-dimensional rotary crane example. The initial position is given by

$$\gamma_0 = \begin{bmatrix} 5\sqrt{3}\,\text{m} \\ 0 \\ -15\,\text{m} \end{bmatrix} \quad \text{at} \quad t = 0 \tag{6.34}$$

and the final position is given by

$$\gamma_f = \begin{bmatrix} -2\,\text{m} \\ 4\sqrt{3}\,\text{m} \\ -13\,\text{m} \end{bmatrix} \quad \text{at} \quad t = 20\,\text{s} \tag{6.35}$$

The initial configuration of the crane system is specified by

$$p = \begin{bmatrix} \frac{5}{2}\sqrt{3}\,\text{m} & 20\,\text{m} & 2.5\,\text{m} & 5\sqrt{3}\,\text{m} & 0 & 2.5\,\text{m} & 5\sqrt{3}\,\text{m} & 0 \end{bmatrix}^T \tag{6.36}$$

at $t_0 = 0$, and the initial load coordinates are given by

$$x = \begin{bmatrix} 5\sqrt{3}\,\mathrm{m} & 0 & -15\,\mathrm{m} \end{bmatrix}^T \tag{6.37}$$

The numerical results for the time step size $\Delta t = 0.01\,\mathrm{s}$ are displayed in Fig. 6.13, in which the coordinates, control inputs and the Lagrange multipliers are presented. The simulated motion of the crane in terms of redundant coordinates is presented in Fig. 6.14 with some snapshots at consecutive points in time.

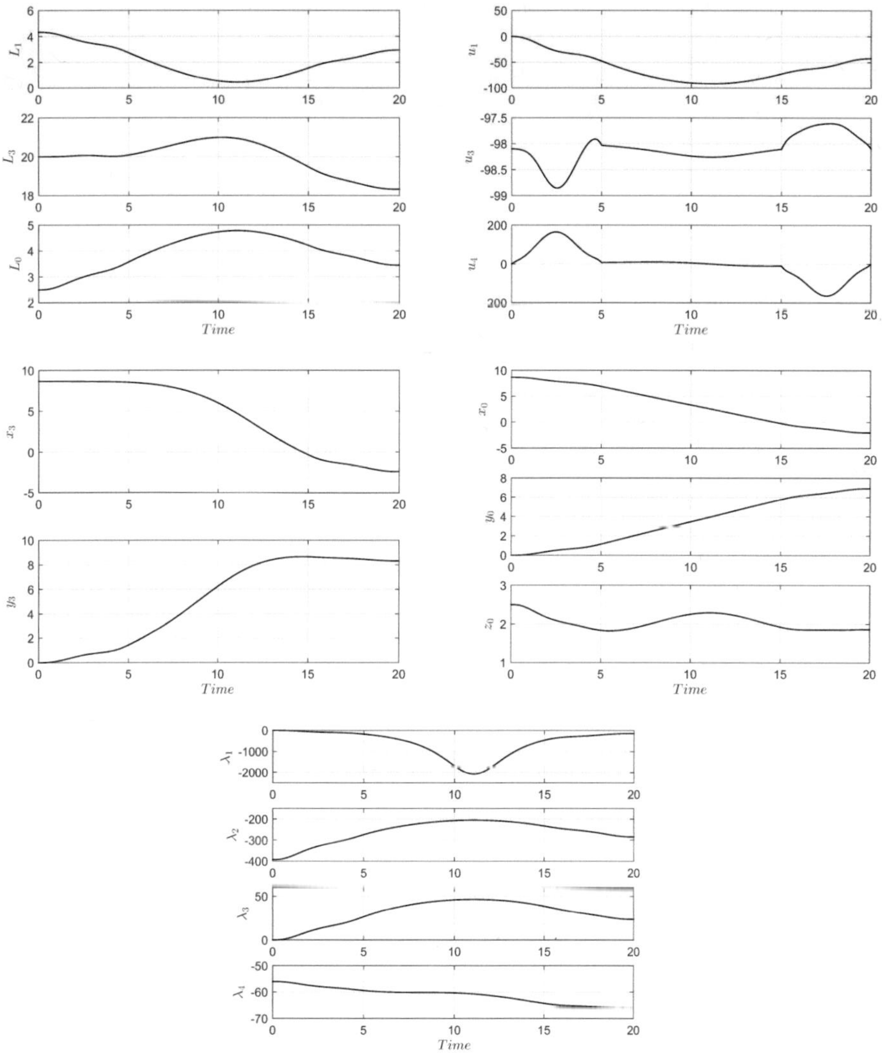

Figure 6.13: Three-dimensional US Navy crane with neglected pulley mass: Numerical results (NUM) obtained with $\Delta t = 0.01\,\mathrm{s}$.

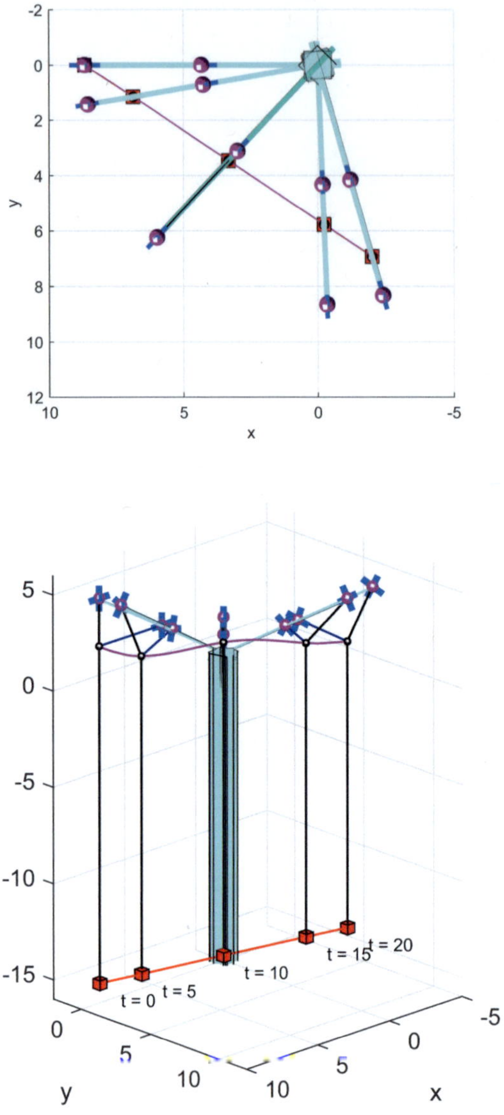

Figure 6.14: Three-dimensional US Navy crane with neglected pulley mass: Snapshots of the load mass and the pulley at specific points in time. Besides the trajectory of pulley and the prescribed trajectory of load mass is shown.

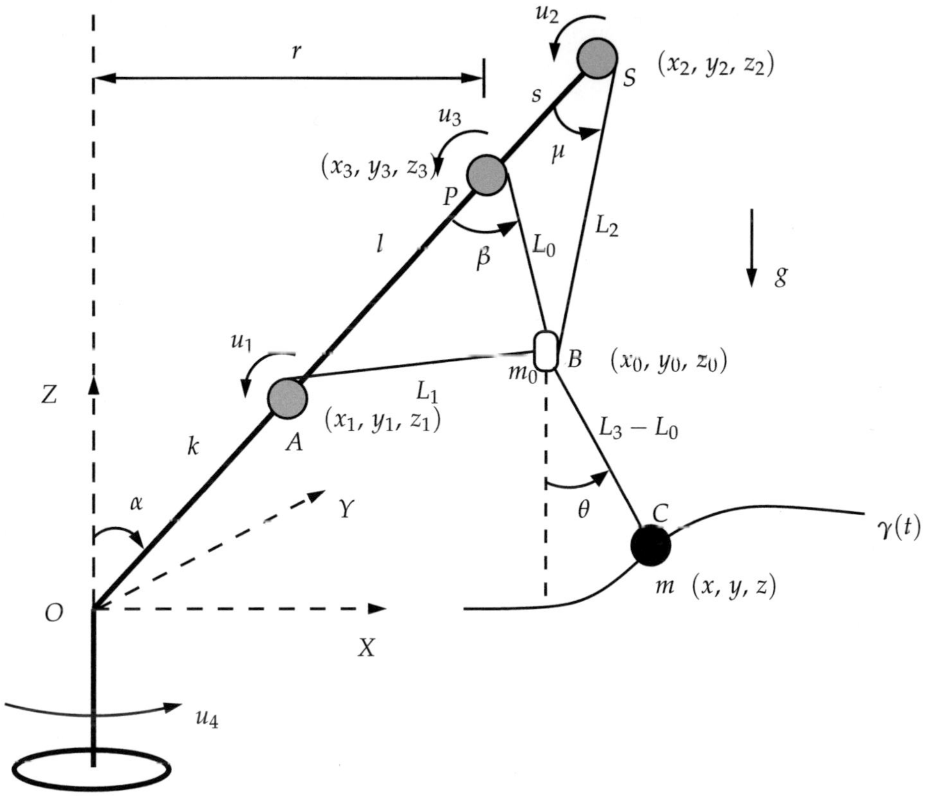

Figure 6.15: The three-dimensional US Navy crane model with nonzero pulley mass ($m_0 > 0$) in terms of $n = 12$ redundant coordinates.

6.2.2 Three-dimensional US Navy crane with nonzero pulley mass

The case, in which the mass of the pulley is nonzero ($m_0 > 0$), is here considered. The suspension cable of variable length L_2 is now needed to pull the pulley.

As shown in Fig. 6.15, the crane is made up of a pole and a system of three cables actuated by three winches and linked by a free pulley. The pole makes a fixed angle α with the vertical rotation axis (Z-axis) of the crane, and is equipped with three winches, one located at point S with Cartesian coordinates (x_2, y_2, z_2), at a fixed distance s from point P, the second located at point P with Cartesian coordinates (x_3, y_3, z_3), at a fixed distance l from point A, and the third located at point A with Cartesian coordinates (x_1, y_1, z_1), at a fixed distance k from the origin O of the inertial reference frame. The first cable of variable length L_3, whose upper part of variable length L_0 makes an angle β with the pole and whose lower part of variable length $L_3 - L_0$ makes an angle θ with the vertical, starts from the winch at point P (radius r_3, moment of inertia J_3, actuating torque u_3), passes through the free pulley located at point B with Cartesian coordinates (x_0, y_0, z_0), and ends up on the load (mass m) located at point C. The second cable of variable length L_1 relates the winch at point A (radius r_1, moment of inertia J_1, actuating torque u_1) to the pulley at point B. The third cable of variable length L_2 connects the winch at point S (radius r_2, moment of inertia J_2, actuating torque u_2) to the free pulley. All the cables are assumed to be massless and unstretchable. It is necessary to use the control input u_4 to rotate the whole crane system in three dimensions. The rotational inertia of the platform is denoted by the mass M. Moreover, r denotes the perpendicular distance between the winch at point P and the rotation axis of the crane.

Redundant coordinates formulation

Similar to the model in the previous subsection, the three-dimensional US Navy crane with nonzero pulley mass can be formulated in terms of $n = 12$ redundant coordinates subjected to $m = 5$ holonomic constraints.

The enlarged set of redundant crane coordinates (see Fig. 6.15) is given by

$$p = \begin{bmatrix} L_1 & L_2 & L_3 & L_0 & x_0 & y_0 & x_3 & y_3 \end{bmatrix}^T \tag{6.38}$$

and by

$$x = \begin{bmatrix} x & y & z & z_0 \end{bmatrix}^T \tag{6.39}$$

The coordinate z_0 of the free pulley is chosen as the fourth flat output due to the introduction of the new control input u_4. Other possible choices for the fourth flat output could be the cable variable $L_3 - L_0$, the coordinate x_0 or y_0 of the free pulley.

The position of the winch at point A is fixed relative to the position of the winch at point P by a parameter $\beta_1 = \frac{1}{2}$, while the position of the winch at point S is fixed relative to the position of the winch at point P by a parameter $\beta_2 = \frac{3}{2}$. The holonomic constraint $\boldsymbol{h}(\boldsymbol{p}) = \boldsymbol{0}$ is given by

$$\boldsymbol{h}(\boldsymbol{p}) = \frac{1}{2}(x_3^2 + y_3^2 - r^2) \tag{6.40}$$

Accordingly, $m_1 = 1$. The above constraint links the parameter r to the position of the winch at point P.

Furthermore, the holonomic constraints $\boldsymbol{\Phi}(\boldsymbol{p}, \boldsymbol{x}) = \mathbf{0}$ are specified by

$$\boldsymbol{\Phi}(\boldsymbol{p}, \boldsymbol{x}) = \begin{bmatrix} \frac{1}{2}\left((x_0 - \beta_1 x_3)^2 + (y_0 - \beta_1 y_3)^2 + (z_0 - \beta_1 z_3)^2 - L_1^2\right) \\ \frac{1}{2}\left((x_0 - \beta_2 x_3)^2 + (y_0 - \beta_2 y_3)^2 + (z_0 - \beta_2 z_3)^2 - L_2^2\right) \\ \frac{1}{2}\left((x_0 - x_3)^2 + (y_0 - y_3)^2 + (z_0 - z_3)^2 - L_0^2\right) \\ \frac{1}{2}\left((x - x_0)^2 + (y - y_0)^2 + (z - z_0)^2 - (L_3 - L_0)^2\right) \end{bmatrix} \tag{6.41}$$

The first constraint links the coordinate L_1 to the position of the pulley and the winch at point A, the second constraint links the coordinate L_2 to the position of the pulley and the winch at point S, the third constraint links the coordinate L_0 to the position of the pulley and the winch at point P, and the last constraint connects the load coordinates to the position of the pulley. Accordingly, $m_2 = 4$.

The total kinetic energy of the crane system can be expressed in terms of redundant coordinates by

$$T = \frac{1}{2}\dot{\boldsymbol{p}} \cdot \boldsymbol{M}_1 \dot{\boldsymbol{p}} + \frac{1}{2}\dot{\boldsymbol{x}} \cdot \boldsymbol{M}_2 \dot{\boldsymbol{x}} \tag{6.42}$$

in which the mass matrices corresponding to the crane coordinates and the extended load coordinates in Equation (6.39) are given by

$$\boldsymbol{M}_1 = \begin{bmatrix} \frac{J_1}{r_1^2} & 0 & 0 & 0 & 0 & 0 & 0 & 0 \\ 0 & \frac{J_2}{r_2^2} & 0 & 0 & 0 & 0 & 0 & 0 \\ 0 & 0 & \frac{J_3}{r_3^2} & 0 & 0 & 0 & 0 & 0 \\ 0 & 0 & 0 & 0 & 0 & 0 & 0 & 0 \\ 0 & 0 & 0 & 0 & m_0 & 0 & 0 & 0 \\ 0 & 0 & 0 & 0 & 0 & m_0 & 0 & 0 \\ 0 & 0 & 0 & 0 & 0 & 0 & M & 0 \\ 0 & 0 & 0 & 0 & 0 & 0 & 0 & M \end{bmatrix}, \quad \boldsymbol{M}_2 = \begin{bmatrix} m & 0 & 0 & 0 \\ 0 & m & 0 & 0 \\ 0 & 0 & m & 0 \\ 0 & 0 & 0 & m_0 \end{bmatrix} \tag{6.43}$$

Further quantities needed in Equations (5.102a)–(5.102d) are given by

$$\boldsymbol{B}_1^T = \begin{bmatrix} \frac{1}{r_1} & 0 & 0 & 0 \\ 0 & \frac{1}{r_2} & 0 & 0 \\ 0 & 0 & \frac{1}{r_3} & 0 \\ 0 & 0 & 0 & 0 \\ 0 & 0 & 0 & 0 \\ 0 & 0 & 0 & 0 \\ 0 & 0 & 0 & -\frac{y_3}{r^2} \\ 0 & 0 & 0 & \frac{x_3}{r^2} \end{bmatrix}, \quad f_1 = \begin{bmatrix} 0 \\ 0 \\ 0 \\ 0 \\ 0 \\ 0 \\ 0 \\ 0 \end{bmatrix}, \quad f_2 = \begin{bmatrix} 0 \\ 0 \\ -mg \\ -m_0 g \end{bmatrix}, \quad \boldsymbol{u} = \begin{bmatrix} u_1 \\ u_2 \\ u_3 \\ u_4 \end{bmatrix} \tag{6.44}$$

and

$$\boldsymbol{H}_1^T = \begin{bmatrix} 0 \\ 0 \\ 0 \\ 0 \\ 0 \\ 0 \\ x_3 \\ y_3 \end{bmatrix}, \quad \boldsymbol{G}_1^T = \begin{bmatrix} -L_1 & 0 & 0 & 0 \\ 0 & -L_2 & 0 & 0 \\ 0 & 0 & 0 & L_0 - L_3 \\ 0 & 0 & -L_0 & L_3 - L_0 \\ x_0 - \beta_1 x_3 & x_0 - \beta_2 x_3 & x_0 - x_3 & x_0 - x \\ y_0 - \beta_1 y_3 & y_0 - \beta_2 y_3 & y_0 - y_3 & y_0 - y \\ -\beta_1(x_0 - \beta_1 x_3) & -\beta_2(x_0 - \beta_2 x_3) & x_3 - x_0 & 0 \\ -\beta_1(y_0 - \beta_1 y_3) & -\beta_2(y_0 - \beta_2 y_3) & y_3 - y_0 & 0 \end{bmatrix} \tag{6.45}$$

and

$$\boldsymbol{G}_2^T = \begin{bmatrix} 0 & 0 & 0 & x - x_0 \\ 0 & 0 & 0 & y - y_0 \\ 0 & 0 & 0 & z - z_0 \\ z_0 - \beta_1 z_3 & z_0 - \beta_2 z_3 & z_0 - z_3 & z_0 - z \end{bmatrix} \tag{6.46}$$

In summary, the three-dimensional US Navy crane with nonzero pulley mass has $f = 7$ degrees of freedom, $a = 4$ control inputs and can be classified as differentially flat system. However, it is very difficult to derive the flatness-based solution as well. The equations of motion are given in detail in Appendix A.5.

Inverse dynamics simulation

In the numerical experiment, the following parameters are applied: $m = 100\,\text{kg}$, $m_0 = 5\,\text{kg}$, $M = 6.4\,\text{kg}$, $J_1 = J_2 = J_3 = 0.1\,\text{kg}\cdot\text{m}^2$, $r_1 = r_2 = r_3 = 0.1\,\text{m}$, $\alpha = \frac{\pi}{3}$, and $s = l = k = 5\,\text{m}$. The prescribed trajectory is obtained in the same way as before. The initial position is given by

$$\gamma_0 = \begin{bmatrix} 5\sqrt{3}\,\mathrm{m} \\ 0 \\ -15\,\mathrm{m} \\ 2.5\,\mathrm{m} \end{bmatrix} \quad \text{at} \quad t = 0 \tag{6.47}$$

and the final position is given by

$$\gamma_f = \begin{bmatrix} -2\,\mathrm{m} \\ 4\sqrt{3}\,\mathrm{m} \\ -13\,\mathrm{m} \\ 1.5\,\mathrm{m} \end{bmatrix} \quad \text{at} \quad t = 20\,\mathrm{s} \tag{6.48}$$

The initial configuration of the crane system is specified by

$$p = \begin{bmatrix} \frac{5}{2}\sqrt{3}\,\mathrm{m} & 6.6144\,\mathrm{m} & 20\,\mathrm{m} & 2.5\,\mathrm{m} & 5\sqrt{3}\,\mathrm{m} & 0 & 5\sqrt{3}\,\mathrm{m} & 0 \end{bmatrix}^T \tag{6.49}$$

at $t_0 = 0$, and the initial load coordinates are given by

$$x = \begin{bmatrix} 5\sqrt{3}\,\mathrm{m} & 0 & -15\,\mathrm{m} & 2.5\,\mathrm{m} \end{bmatrix}^T \tag{6.50}$$

The numerical results for the time step size $\Delta t = 0.1\,\mathrm{s}$ are displayed in Fig. 6.16, in which the coordinates, control inputs and the Lagrange multipliers are presented. The simulated motion of the crane in terms of redundant coordinates is presented in Fig. 6.17 with some snapshots at consecutive points in time.

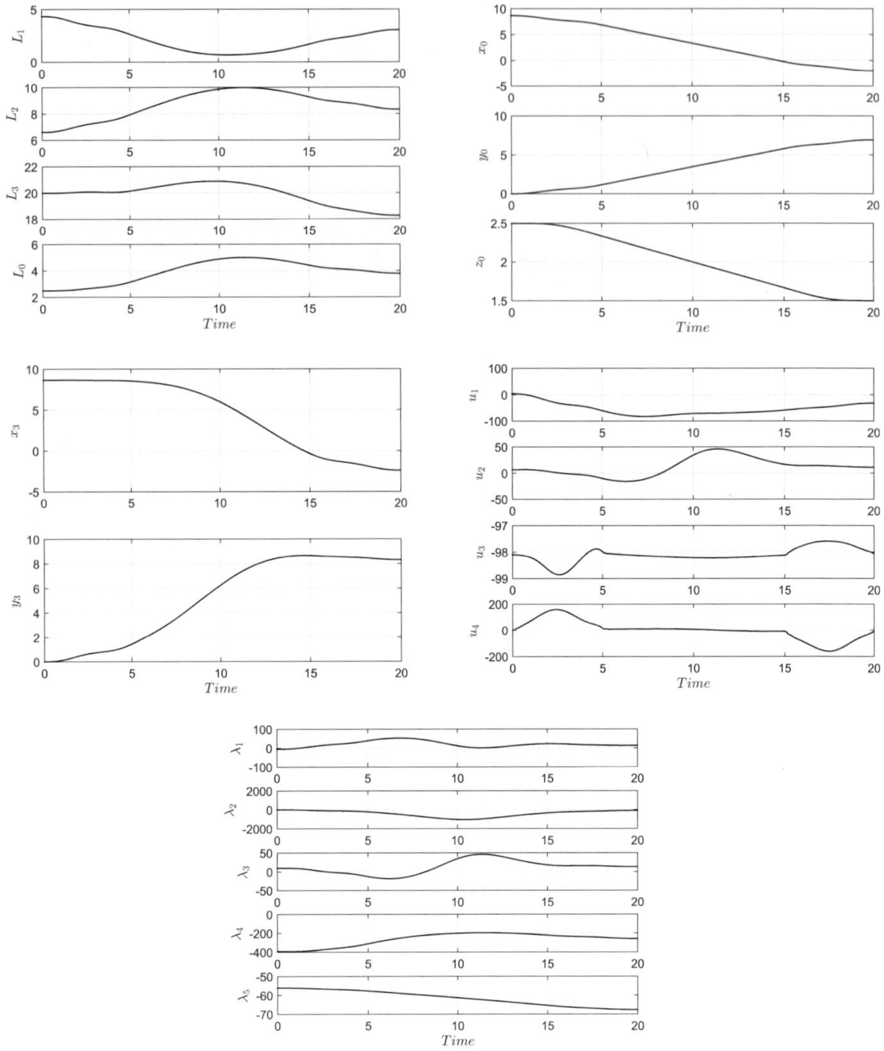

Figure 6.16: Three-dimensional US Navy crane with nonzero pulley mass: Numerical results (NUM) obtained with $\Delta t = 0.1\,\mathrm{s}$.

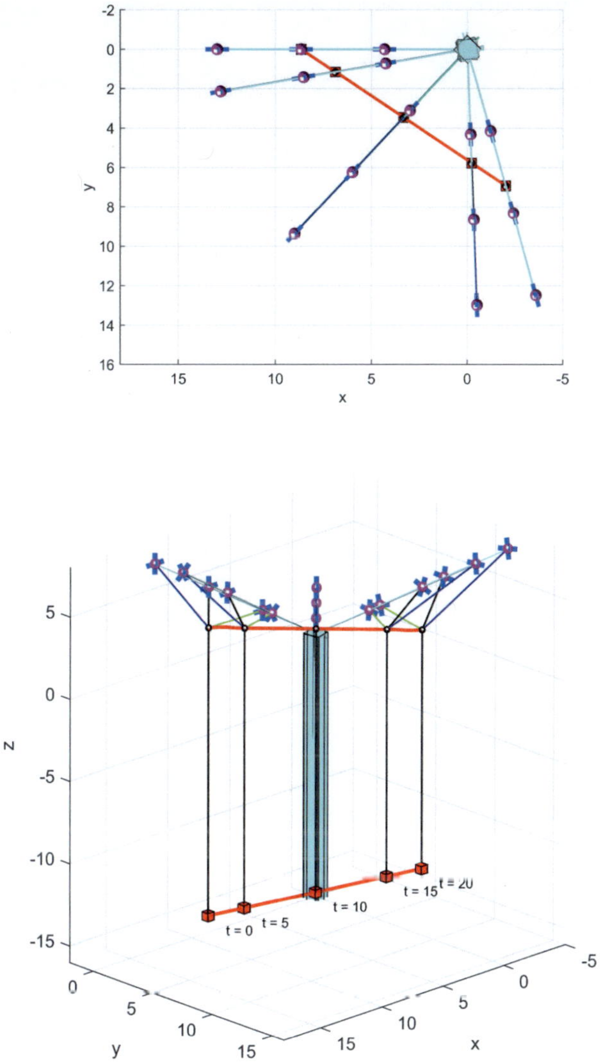

Figure 6.17: Three-dimensional US Navy crane with nonzero pulley mass: Snapshots of the load mass and the pulley at specific points in time. Besides the trajectory of pulley and the prescribed trajectory of load mass is shown.

6.3 Cable suspension manipulator

Cable suspension manipulators support a payload platform in space by several spatially arranged cables with computer-controlled winches. The winches are mounted on movable trolleys and are responsible for regulating the lengths of the cables. Compared to the crane models considered before, it is possible to control not only the translational motion of the payload but also its orientation in order to perform, for example, assembly tasks. Therefore, cable suspension manipulators combine the ability of cranes to support heavy payloads in a large workspace with the dexterity of robot manipulators [53].

Cable suspension manipulators can be classified as kinematically/statically determined or kinematically/statically undetermined. The elaborate description can be found in [53]. Here a prototype of the three-cable suspension manipulator (CABLEV) is considered (see Fig. 6.18), which has been developed at the University of Rostock and treated in nonlinear trajectory tracking control problems (see [53, 52, 70]).

The CABLEV manipulator under consideration is kinematically undetermined because the platform is (finitely or infinitesimally) movable while the cable lengths are kept constant. It implies that the payload platform may perform sway motions with three degrees of freedom. The payload platform is suspended by three cables with three winches mounted on trolleys that move themselves on a gantry. It is also possible for the gantry to move on the rails. Applications for such systems are, for example, precise handling and assembling large and heavy components on construction sites or on shipyards [53].

Since the kinematically undetermined cable suspension manipulator can be classified as an underactuated mechanical system, the platform can not be controlled like the end-effector of a conventional robot by inverse dynamics control. The position of the platform is not uniquely determined by the robot (crane) coordinates of the trolleys and winches. In contrast to the flatness-based feedforward and nonlinear feedback control applied in [53], the new approach, index reduction by minimal extension, is applied to the dynamic model of the cable suspension manipulator to obtain the feedforward control law in the case of vanishing disturbances. In addition, a closed-loop control strategy with feedback of actual errors in load position and orientation provides stable tracking of required reference load trajectory in presence of pertubations.

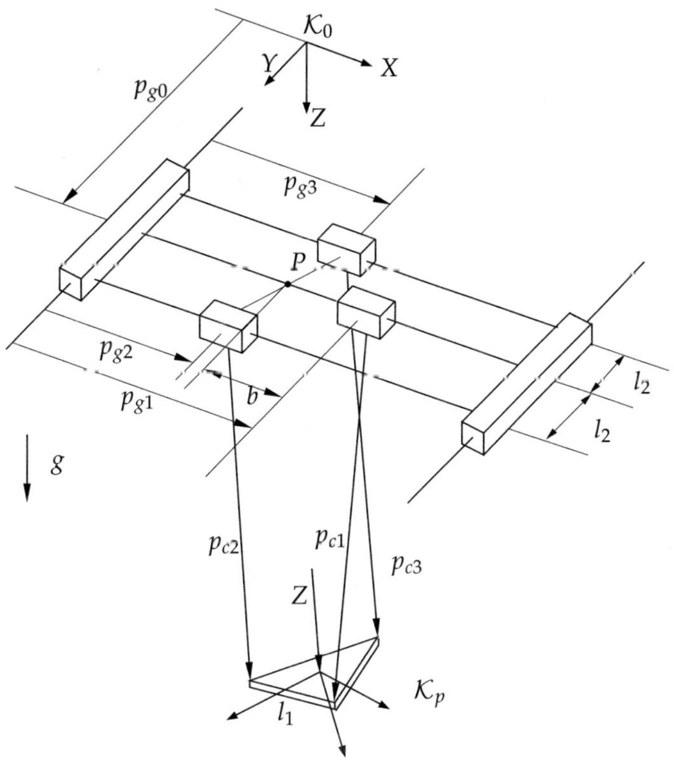

Figure 6.18: The three-cable suspension manipulator model (CABLEV).

6.3.1 Redundant coordinates

The dynamic model of CABLEV can be formulated in terms of $n = 13$ redundant coordinates subjected to $m = 3$ holonomic constraints. The set of redundant crane coordinates $\boldsymbol{p} \in \mathbb{R}^7$ (see Fig. 6.18) is given by

$$\boldsymbol{p} = \begin{bmatrix} \boldsymbol{p}_g \\ \boldsymbol{p}_c \end{bmatrix} \tag{6.51}$$

with gantry coordinates

$$\boldsymbol{p}_g = \begin{bmatrix} p_{g0} & p_{g1} & p_{g2} & p_{g3} \end{bmatrix}^T \tag{6.52}$$

and cable coordinates

$$\boldsymbol{p}_c = \begin{bmatrix} p_{c1} & p_{c2} & p_{c3} \end{bmatrix}^T \tag{6.53}$$

Here, p_{g0} denotes the position of the gantry on the rails. The displacements of three trolleys on the movable gantry are described by p_{gi} $(i = 1, 2, 3)$. Moreover, the coordinates p_{ci} $(i = 1, 2, 3)$, are the lengths of three cables connecting the platform with the winches.

The platform coordinates $\boldsymbol{x} \in \mathbb{R}^6$ are expressed by

$$\boldsymbol{x} = \begin{bmatrix} \boldsymbol{r} \\ \boldsymbol{\varphi} \end{bmatrix} \tag{6.54}$$

with the position vector

$$\boldsymbol{r} = \begin{bmatrix} r_x & r_y & r_z \end{bmatrix}^T \tag{6.55}$$

and the angles

$$\boldsymbol{\varphi} = \begin{bmatrix} \varphi_1 & \varphi_2 & \varphi_3 \end{bmatrix}^T \tag{6.56}$$

Here, the spatial position and orientation of the platform-fixed frame $\mathcal{K}_p$ relative to the inertial fram $\mathcal{K}_0$ are described by three Cartesian coordinates r_x, r_y, r_z of the origin of $\mathcal{K}_p$ and three Bryant angles $\varphi_1, \varphi_2, \varphi_3$. Note that the origin of the body-fixed frame $\mathcal{K}_p$ coincides with the center of mass of the payload platform.

The spatial velocity of the platform relative to $\mathcal{K}_0$ is given by the twist

$$\boldsymbol{t} = \begin{bmatrix} \boldsymbol{v} \\ \boldsymbol{\omega} \end{bmatrix} \tag{6.57}$$

with the velocity vector

$$\boldsymbol{v} = \begin{bmatrix} \dot{r}_x & \dot{r}_y & \dot{r}_z \end{bmatrix}^T \tag{6.58}$$

and the angular velocities

$$\boldsymbol{\omega} = \begin{bmatrix} \omega_x & \omega_y & \omega_z \end{bmatrix}^T \tag{6.59}$$

Here, $\dot{r}_x, \dot{r}_y, \dot{r}_z$ are the translational velocity coordinates of the origin of $\mathcal{K}_p$, and ω_x, ω_y, ω_z are the coordinates of the angular velocity $\boldsymbol{\omega}$ in $\mathcal{K}_0$. The relation between the derivative $\dot{\boldsymbol{x}}$ and the twist $\boldsymbol{t}$ is then given by

$$\dot{\boldsymbol{x}} = \boldsymbol{H}(\boldsymbol{x})\,\boldsymbol{t} \tag{6.60a}$$

$$\begin{bmatrix} \dot{\boldsymbol{r}} \\ \dot{\boldsymbol{\varphi}} \end{bmatrix} = \begin{bmatrix} \boldsymbol{I}_3 & \boldsymbol{0} \\ \boldsymbol{0} & \boldsymbol{H}_\omega(\boldsymbol{\varphi}) \end{bmatrix} \begin{bmatrix} \boldsymbol{v} \\ \boldsymbol{\omega} \end{bmatrix} \tag{6.60b}$$

with the expression

$$H(x) = \begin{bmatrix} I_3 & 0 \\ 0 & H_\omega(\varphi) \end{bmatrix} \tag{6.61}$$

The kinematic differential equation related to the Bryant angles can be found from the second matrix equation in (6.60b). It is

$$\dot{\varphi} = H_\omega(\varphi)\,\omega \tag{6.62a}$$

$$\begin{bmatrix} \dot\varphi_1 \\ \dot\varphi_2 \\ \dot\varphi_3 \end{bmatrix} = \frac{1}{\cos\varphi_2} \begin{bmatrix} \cos\varphi_2 & \sin\varphi_1 \sin\varphi_2 & -\cos\varphi_1 \sin\varphi_2 \\ 0 & \cos\varphi_1 \cos\varphi_2 & \sin\varphi_1 \cos\varphi_2 \\ 0 & -\sin\varphi_1 & \cos\varphi_1 \end{bmatrix} \begin{bmatrix} \omega_x \\ \omega_y \\ \omega_z \end{bmatrix} \tag{6.62b}$$

with the expression

$$H_\omega(\varphi) = \frac{1}{\cos\varphi_2} \begin{bmatrix} \cos\varphi_2 & \sin\varphi_1 \sin\varphi_2 & -\cos\varphi_1 \sin\varphi_2 \\ 0 & \cos\varphi_1 \cos\varphi_2 & \sin\varphi_1 \cos\varphi_2 \\ 0 & -\sin\varphi_1 & \cos\varphi_1 \end{bmatrix} \tag{6.63}$$

Furthermore, the inverse kinematic differential equation is

$$t = H^{-1}(x)\,\dot{x} \tag{6.64}$$

The same notation as in [53] is also used here. For the convenience of expressing the constraint Jacobian matrix, the quasi-coordinates s, that exist only as differentials, can be defined. That implies

$$\dot{s} = t = \begin{bmatrix} v \\ \omega \end{bmatrix} \tag{6.65}$$

With 3 suspending cables the system is kinematically undetermined, i.e. the payload platform can perform sway motions with $f = 6 - 3 = 3$ degrees of freedom [53]. The sway coordinates x_1, that describe the sway motions, are chosen as

$$x_1 = \begin{bmatrix} r_x & r_y & \varphi_3 \end{bmatrix}^T \tag{6.66}$$

and the remaining coordinates read

$$x_2 = \begin{bmatrix} r_z & \varphi_1 & \varphi_2 \end{bmatrix} \tag{6.67}$$

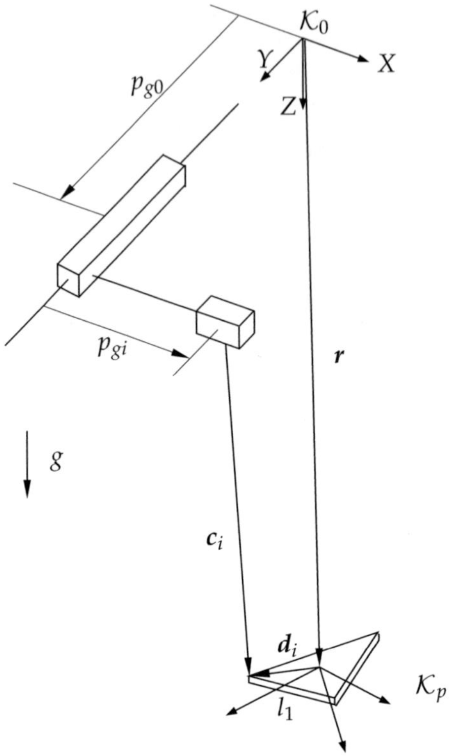

Figure 6.19: Schematic of one holonomic constraint.

6.3.2 Constraints

The relations between robot coordinates $\boldsymbol{p}$ and platform coordinates $\boldsymbol{x}$ in Fig. 6.19 can be described by the holonomic constraints $\boldsymbol{\Phi}(\boldsymbol{p},\boldsymbol{x}) = \boldsymbol{0}$, that can be written as

$$\Phi_i(\boldsymbol{p},\boldsymbol{x}) = \frac{1}{2}\left(\boldsymbol{c}_i^T(\boldsymbol{p},\boldsymbol{x})\boldsymbol{c}_i(\boldsymbol{p},\boldsymbol{x}) - p_{ci}^2\right), \qquad i = 1,2,3 \tag{6.68}$$

The cable vectors $\boldsymbol{c}_i$ are expressed in $\mathcal{K}_0$ by

$$\boldsymbol{c}_1 = \boldsymbol{r} + \boldsymbol{d}_1 - p_{g0}\,\boldsymbol{e}_y - p_{g1}\,\boldsymbol{e}_x \tag{6.69a}$$
$$\boldsymbol{c}_2 = \boldsymbol{r} + \boldsymbol{d}_2 - (p_{g0} + l_2)\,\boldsymbol{e}_y - p_{g2}\,\boldsymbol{e}_x \tag{6.69b}$$
$$\boldsymbol{c}_3 = \boldsymbol{r} + \boldsymbol{d}_3 - (p_{g0} - l_2)\,\boldsymbol{e}_y - p_{g3}\,\boldsymbol{e}_x \tag{6.69c}$$

Here, l_2 denotes the distance between the rails (see Fig. 6.18), and the unit vectors $\boldsymbol{e}_x$ and $\boldsymbol{e}_y$ of $\mathcal{K}_0$ are expressed by $\boldsymbol{e}_x = \begin{bmatrix} 1 & 0 & 0 \end{bmatrix}^T$ and $\boldsymbol{e}_y = \begin{bmatrix} 0 & 1 & 0 \end{bmatrix}^T$. It is assumed that the platform is an equilateral triangle with the side length l_1. The body-fixed vectors $\boldsymbol{d}_i$ point from the origin of $\mathcal{K}_p$ to each corner point of the equilateral triangle. The constant coordinates of the body-fixed vectors $\boldsymbol{d}_i$ can be expressed in $\mathcal{K}_p$ by

$$\boldsymbol{X}_1 = \begin{bmatrix} \frac{1}{3}\sqrt{3}l_1 & 0 & 0 \end{bmatrix}^T \tag{6.70a}$$

$$\boldsymbol{X}_2 = \begin{bmatrix} -\frac{1}{6}\sqrt{3}l_1 & \frac{1}{2}l_1 & 0 \end{bmatrix}^T \tag{6.70b}$$

$$\boldsymbol{X}_3 = \begin{bmatrix} -\frac{1}{6}\sqrt{3}l_1 & -\frac{1}{2}l_1 & 0 \end{bmatrix}^T \tag{6.70c}$$

The body-fixed vectors $\boldsymbol{d}_i$ expressed in $\mathcal{K}_0$ can be given by

$$\boldsymbol{d}_i = \boldsymbol{R}(\boldsymbol{\varphi})\boldsymbol{X}_i, \qquad i = 1, 2, 3 \tag{6.71}$$

with the transformation (rotation) matrix $\boldsymbol{R}(\boldsymbol{\varphi})$ from the body-fixed frame $\mathcal{K}_p$ to the inertial frame $\mathcal{K}_0$.

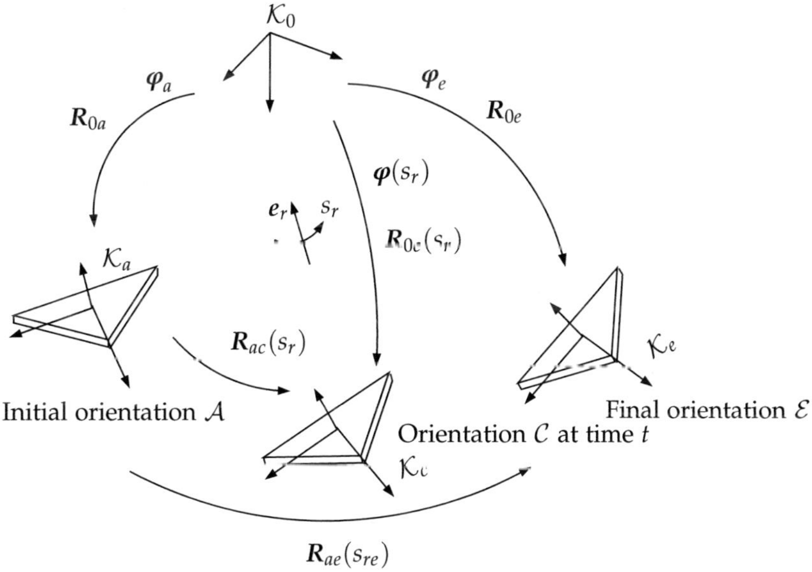

Figure 6.20: Rotational motion of the payload platform about the axis of rotation $\boldsymbol{e}_r$ with the rotation angle s_{re} between the initial and final orientation.

For example, as depicted in Fig. 6.20, the rotation matrix $\boldsymbol{R}(\boldsymbol{\varphi})$ from the current orientation $\mathcal{K}_c$ at time t to the inertial frame $\mathcal{K}_0$ reads

$$\boldsymbol{R}_{0c}(s_r) = \boldsymbol{R}_{0a}(\boldsymbol{\varphi}_a)\boldsymbol{R}_{ac}(\boldsymbol{e}_r, s_r) \tag{6.72}$$

The rotation matrix $\boldsymbol{R}_{0a}$ is described by application of three Bryant angles $\boldsymbol{\varphi}_a$. The rotation matrix $\boldsymbol{R}_{ac}$ is described by using Rodrigues formula. In this connection, $\boldsymbol{e}_r$ is the unit vector of the rotation axis between the initial orientation $\mathcal{K}_a$ and the final orientation $\mathcal{K}_e$. Moreover, s_r is the rotation angle of the frame $\mathcal{K}_c$ at time t about the rotation axis $\boldsymbol{e}_r$. Similarly, the rotation matrix $\boldsymbol{R}_{0e}$ can be described by application of three Bryant angles $\boldsymbol{\varphi}_e$. In order to calculate the constraint Jacobian matrices associated to the holonomic constraints (6.68), differentiating the cable vectors (6.69) with respect to time yields

$$\dot{\boldsymbol{c}}_i = \dot{\boldsymbol{r}} + \dot{\boldsymbol{d}}_i - \dot{p}_{g0}\,\boldsymbol{e}_y - \dot{p}_{gi}\,\boldsymbol{e}_x \qquad i = 1,2,3 \tag{6.73}$$

Similar to the director velocities introduced in Subsection 2.3.2, the first time derivative of the body-fixed vectors $\boldsymbol{d}_i$ can be calculated through the angular velocity of the platform $\boldsymbol{\omega}$ [10]. Thus

$$\dot{\boldsymbol{d}}_i = \boldsymbol{\omega} \times \boldsymbol{d}_i = -\boldsymbol{d}_i \times \boldsymbol{\omega} = -\widehat{\boldsymbol{d}}_i\boldsymbol{\omega} \tag{6.74}$$

with the skew-symmetric matrix

$$\widehat{\boldsymbol{d}}_i = \begin{bmatrix} 0 & -d_{3_i} & d_{2_i} \\ d_{3_i} & 0 & -d_{1_i} \\ -d_{2_i} & d_{1_i} & 0 \end{bmatrix} \tag{6.75}$$

Then differentiating the constraint equations (6.68) with respect to time leads to

$$\dot{\Phi}_i = \boldsymbol{c}_i^T\dot{\boldsymbol{c}}_i - p_{ci}\,\dot{p}_{ci} = 0 \qquad i = 1,2,3 \tag{6.76}$$

By inserting Equation (6.73) with (6.74) into (6.76), the constraint equation at the velocity level can be written as

$$\dot{\boldsymbol{\Phi}} = \boldsymbol{G}_s(\boldsymbol{p},\boldsymbol{x})\begin{bmatrix} \boldsymbol{v} \\ \boldsymbol{\omega} \end{bmatrix} + \boldsymbol{G}_p(\boldsymbol{p},\boldsymbol{x})\begin{bmatrix} \dot{\boldsymbol{p}}_g \\ \dot{\boldsymbol{p}}_c \end{bmatrix} = \boldsymbol{0} \tag{6.77}$$

with the constraint Jacobian matrix

$$G_s = \begin{bmatrix} c_1^T & -c_1^T \widehat{d}_1 \\ c_2^T & -c_2^T \widehat{d}_2 \\ c_3^T & -c_3^T \widehat{d}_3 \end{bmatrix} \tag{6.78}$$

and

$$G_p = -\begin{bmatrix} c_1^T e_y & c_1^T e_x & 0 & 0 & p_{c1} & 0 & 0 \\ c_2^T e_y & 0 & c_2^T e_x & 0 & 0 & p_{c2} & 0 \\ c_3^T e_y & 0 & 0 & c_3^T e_x & 0 & 0 & p_{c3} \end{bmatrix} \tag{6.79}$$

The above Equation (6.77) can be rewritten in x, $\dot{x}$, p and p by inserting the inverse kinematic Equation (6.64). That is

$$\dot{\Phi} = G_s(p, x) H^{-1} \begin{bmatrix} v \\ \dot{\varphi} \end{bmatrix} + G_p(p, x) \begin{bmatrix} \dot{p}_g \\ \dot{p}_c \end{bmatrix} = \mathbf{0} \tag{6.80}$$

Kinematic redundancy and flat outputs

The system of CABLEV is kinematically redundant, since it has seven control inputs corresponding to the robot coordinates p and six load coordinates of the platform x. Accordingly, a seventh flat output x_0 can be defined as the residual of the implicit control constraint

$$x_0 = \Phi_0(p) = p_{g1} - \frac{1}{2}(p_{g2} + p_{g3}) - b = 0 \tag{6.81}$$

It implies that the distance b between the inner trolley and the intersection point P (between the line through the outer trolleys and the intermediate rail) is constant, for example, $b = \sqrt{3}\, l_2$ (see Fig. 6.18). This constraint makes sure that the shape of the triangle, whose vertices are the three trolleys, is constrained. To summarize, the flat outputs are composed of the load coordinates x and the additional output variable x_0,

$$x_f = \begin{bmatrix} x \\ x_0 \end{bmatrix} \tag{6.82}$$

The flat outputs at the velocity level are expressed by the time derivative of quasi-coordinates $\dot{s}$ and $\dot{x}_0$,

$$\dot{s}_f = \begin{bmatrix} \dot{s} \\ \dot{x}_0 \end{bmatrix} \tag{6.83}$$

Correspondingly, the flat outputs at the acceleration level are defined as

$$\ddot{\mathbf{s}}_f = \begin{bmatrix} \ddot{\mathbf{s}} \\ \ddot{x}_0 \end{bmatrix} \tag{6.84}$$

These constraint conditions will be used in the index reduction by minimal extension method in the sequel.

6.3.3 Dynamic equations

The dynamic equations of CABLEV consist of the equations of the drive system including the gantry, trolleys and winches, the equations of the payload platform, and the holonomic and servo constraint equations. The governing equations assume the form of differential-algebraic equations with high index, since CABLEV is an underactuated mechanical system with $f = 10$ degrees of freedom and $a = 7$ control inputs.

Dynamics of the drive system

The equations of motion of the drive system are formulated in terms of the robot coordinates $\boldsymbol{p}$,

$$\boldsymbol{M}_1 \ddot{\boldsymbol{p}} = \boldsymbol{f}_1(\boldsymbol{p}, \dot{\boldsymbol{p}}) + \boldsymbol{B}_1^T \boldsymbol{u} + \boldsymbol{G}_p^T(\boldsymbol{p}, \boldsymbol{x})\boldsymbol{\lambda} \tag{6.85}$$

with the mass matrix $\boldsymbol{M}_1 \in \mathbb{R}^{7,7}$, the force vector $\boldsymbol{f}_1 \in \mathbb{R}^7$ and the input transformation vector $\boldsymbol{B}_1 \in \mathbb{R}^{7,7}$,

$$\boldsymbol{M}_1 = \begin{bmatrix} m_0 & 0 & 0 & 0 & 0 & 0 & 0 \\ 0 & m_1 & 0 & 0 & 0 & 0 & 0 \\ 0 & 0 & m_2 & 0 & 0 & 0 & 0 \\ 0 & 0 & 0 & m_3 & 0 & 0 & 0 \\ 0 & 0 & 0 & 0 & \frac{J_1}{r_1^2} & 0 & 0 \\ 0 & 0 & 0 & 0 & 0 & \frac{J_2}{r_2^2} & 0 \\ 0 & 0 & 0 & 0 & 0 & 0 & \frac{J_3}{r_3^2} \end{bmatrix}, \quad \boldsymbol{f}_1 = \begin{bmatrix} 0 \\ 0 \\ 0 \\ 0 \\ 0 \\ 0 \\ 0 \end{bmatrix}, \quad \boldsymbol{B}_1 = \begin{bmatrix} 1 & 0 & 0 & 0 & 0 & 0 & 0 \\ 0 & 1 & 0 & 0 & 0 & 0 & 0 \\ 0 & 0 & 1 & 0 & 0 & 0 & 0 \\ 0 & 0 & 0 & 1 & 0 & 0 & 0 \\ 0 & 0 & 0 & 0 & \frac{1}{r_1} & 0 & 0 \\ 0 & 0 & 0 & 0 & 0 & \frac{1}{r_2} & 0 \\ 0 & 0 & 0 & 0 & 0 & 0 & \frac{1}{r_3} \end{bmatrix} \tag{6.86}$$

The gantry and trolley mass are denoted by m_0 and m_i, and the radius and moment of inertia of the winch are given by r_i and J_i $(i = 1, 2, 3)$.

The control forces are given by

$$u = \begin{bmatrix} u_g \\ u_c \end{bmatrix}, \quad u_g = \begin{bmatrix} u_{g0} & u_{g1} & u_{g2} & u_{g3} \end{bmatrix}^T, \quad u_c = \begin{bmatrix} u_{c1} & u_{c2} & u_{c3} \end{bmatrix}^T \tag{6.87}$$

with the gantry force u_{g0}, the trolley forces u_{gi} and the winch torques u_{ci} ($i = 1, 2, 3$). The Lagrange multipliers are described by $\boldsymbol{\lambda} = \begin{bmatrix} \lambda_1 & \lambda_2 & \lambda_3 \end{bmatrix}^T$.

Dynamics of the payload platform

The classical Newton-Euler equations for rigid bodies can be applied to derive the governing equations of the payload platform. Thus they are expressed by

$$\boldsymbol{M}_2 \ddot{\boldsymbol{s}} = \boldsymbol{f}_2 + \boldsymbol{G}_s^T(\boldsymbol{p}, \boldsymbol{x})\boldsymbol{\lambda} \tag{6.88}$$

with the mass matrix $\boldsymbol{M}_2 \in \mathbb{R}^{6,6}$, the generalized applied and gyroscopic forces $\boldsymbol{f}_2$,

$$\boldsymbol{M}_2 = \begin{bmatrix} m\boldsymbol{I}_3 & \boldsymbol{0} \\ \boldsymbol{0} & \boldsymbol{J} \end{bmatrix}, \quad \boldsymbol{f}_2 = \begin{bmatrix} m\boldsymbol{g} \\ -\widehat{\boldsymbol{\omega}}\boldsymbol{J}\boldsymbol{\omega} \end{bmatrix} \tag{6.89}$$

and the other terms

$$\boldsymbol{J} = \begin{bmatrix} J_x & 0 & 0 \\ 0 & J_y & 0 \\ 0 & 0 & J_z \end{bmatrix}, \quad \widehat{\boldsymbol{\omega}}\boldsymbol{J}\boldsymbol{\omega} = \begin{bmatrix} (J_z - J_y)\,\omega_y\,\omega_z \\ (J_x - J_z)\,\omega_z\,\omega_x \\ (J_y - J_x)\,\omega_x\,\omega_y \end{bmatrix}, \quad \boldsymbol{g} = \begin{bmatrix} 0 \\ 0 \\ g \end{bmatrix} \tag{6.90}$$

Here, m denotes the mass of the platform, and J_x, J_y, J_z are the principal mass moments of inertia with respect to the center of mass of the equilateral triangle platform represented in coordinates of $\mathcal{K}_0$, i.e. $\boldsymbol{J} = \boldsymbol{R}\,\boldsymbol{J}^p\,\boldsymbol{R}^T$. The inertia tensor with respect to the center of the platform is represented in the body-fixed principal axes system by

$$\boldsymbol{J}^p = \begin{bmatrix} J_x & 0 & 0 \\ 0 & J_y & 0 \\ 0 & 0 & J_z \end{bmatrix} \tag{6.91}$$

Next differentiating the inverse kinematic differential Equation (6.64) with respect to time gives

$$\ddot{\boldsymbol{s}} = \boldsymbol{H}^{-1}\ddot{\boldsymbol{x}} + \dot{\boldsymbol{H}}^{-1}\dot{\boldsymbol{x}} \tag{6.92}$$

Inserting the above Equation (6.92) into (6.88) yields the alternative form of the dynamic equations of the platform

$$M_2 H^{-1}\ddot{x} + M_2 \dot{H}^{-1}\dot{x} = f_2 + G_s^T(p, x)\lambda \tag{6.93}$$

6.3.4 Motion planning

During the operation of CABLEV, the payload platform is required to move from an initial position and/or orientation to a desired final destination and/or orientation in its working space along a trajectory. This procedure needs motion planning for the position of the center of mass of the platform and/or for the spatial orientation of the rigid platform.

For the translational motion of the payload platform, the trajectory of the center of mass can be prescribed by using a reference function $c(t)$ in a similar manner as before. This provides a rest-to-rest maneuver, which can be divided into three phases: the acceleration, steady velocity and deceleration phase.

In Fig. 6.20, for the rotational motion of the payload platform, the Bryant angles φ can be employed to describe the motion of rotations between the initial orientation φ_a and the final orientation φ_e.

In addition, the rotational motion of the payload platform between the initial and final orientation can be represented by the rotation motion around a space-fixed vector e_r, which comes from the origin of the body-fixed frame $\mathcal{K}_p$, with the angle of rotation s_{re}. Thus, the rotation matrix can be calculated by the given axis of rotation e_r with angle of rotation s_{re} via the Rodrigues formula. The conversion between the two formalisms of rotation is necessary for motion planning of rotation in the numerical example at hand.

Given φ_a and φ_e, the rotation matrix between the initial and final orientation is calculated relative to $\mathcal{K}_0$ by

$$R_{ae}(\varphi_a, \varphi_e) = R_{0a}^T(\varphi_a) R_{0e}(\varphi_e) \tag{6.94}$$

By using Rodrigues formula [52, 70], the rotation matrix $R_{ac}(e_r, s_{re})$ is given by

$$R_{ae}(e_r, s_{re}) = \cos(s_{re}) I_3 + \sin(s_{re})\widehat{e}_r + (1 - \cos(s_{re}))\, e_r e_r^T \tag{6.95}$$

The rotation matrix in Equation (6.94) and (6.95) are both equal to each other and the equality leads to the computation of the angle of rotation s_{re}[1] [92]

$$s_{re} = \arccos\left(\frac{1}{2}\left(\boldsymbol{R}_{ae}(1,1) + \boldsymbol{R}_{ae}(2,2) + \boldsymbol{R}_{ae}(3,3) - 1\right)\right) \tag{6.96}$$

and the vector of the rotation axis $\boldsymbol{e}_r$ [92]

$$\boldsymbol{e}_r = \frac{1}{\sin s_{re}} \begin{bmatrix} \boldsymbol{R}_{ae}(3,2) - \boldsymbol{R}_{ae}(2,3) \\ \boldsymbol{R}_{ae}(1,3) - \boldsymbol{R}_{ae}(3,1) \\ \boldsymbol{R}_{ae}(2,1) - \boldsymbol{R}_{ae}(1,2) \end{bmatrix} \tag{6.97}$$

where $\boldsymbol{R}_{ae}(i,j)$ denotes the element at the i-th row and j-th column of the rotation matrix $\boldsymbol{R}_{ae}(\boldsymbol{\varphi}_a, \boldsymbol{\varphi}_e)$. The rotation matrix $\boldsymbol{R}_{0c}$ from $\mathcal{K}_0$ to the current orientation $\mathcal{C}$ can now be calculated through Equation (6.72).

Then the Bryant angles corresponding to the current orientation $\mathcal{C}$ are given by [92]

$$\boldsymbol{\varphi}(s_r) = \begin{bmatrix} \varphi_1(s_r) \\ \varphi_2(s_r) \\ \varphi_3(s_r) \end{bmatrix} \tag{6.98}$$

with the components

$$\varphi_1(s_r) = \arccos\left(\frac{\boldsymbol{R}_{0c}(3,3)}{\cos\varphi_2}\right) \tag{6.99a}$$

$$\varphi_2(s_r) = \arccos\left(\sqrt{1 - \boldsymbol{R}_{0c}^2(1,3)}\right) \tag{6.99b}$$

$$\varphi_3(s_r) = \arccos\left(\frac{\boldsymbol{R}_{0c}(1,1)}{\cos\varphi_2}\right) \tag{6.99c}$$

Similar to the prescribed translational motion, the desired trajectory of rotational motion of the platform is specified by

$$\gamma_s(t) = \gamma_{s0} + (\gamma_{sf} - \gamma_{s0})c(t) \tag{6.100}$$

with $\gamma_{s0} = 0$ at t_0 and $\gamma_{sf} = s_{re}$ at t_f.

[1] Note that the angle of rotation s_{re} belongs to the interval $0 < s_{re} < 180°$.

The reference function $c(t)$ is composed of three phases,

$$c(t) = \begin{cases} c_I(t) & \text{for } 0 \leqslant t < 5\,\text{s} \\ c_{II}(t) & \text{for } 5\,\text{s} \leqslant t < 15\,\text{s} \\ c_{III}(t) & \text{for } 15\,\text{s} \leqslant t \leqslant 20\,\text{s} \end{cases}$$

with each phase

$$\begin{aligned} c_I(t) &= \frac{1}{\tau - \tau_0}\left(-\frac{5t^8}{2\tau_0{}^7} + \frac{10t^7}{\tau_0{}^6} - \frac{14t^6}{2\tau_0{}^5} + \frac{7t^5}{2\tau_0{}^4}\right) \\ c_{II}(t) &= \frac{1}{\tau - \tau_0}\left(t - \frac{\tau_0}{2}\right) \\ c_{III}(t) &= 1 + \frac{1}{\tau - \tau_0}\left(-\frac{5(\tau - t)^8}{2\tau_0{}^7} + \frac{10(\tau - t)^7}{\tau_0{}^6} - \frac{14(\tau - t)^6}{2\tau_0{}^5} + \frac{7(\tau - t)^5}{2\tau_0{}^4}\right) \end{aligned}$$

where $\tau = t_f - t_0$, and τ_0 is the acceleration/deceleration time.

The Bryant angles corresponding to $s_r(t) = \gamma_s(t)$ can be computed through Equation (6.72) and Equations (6.99a)–(6.99c). This provides the prescribed Bryant angles $\gamma_\varphi(t)$ in Equation (6.103h).

Accordingly, the angular velocity about the space-fixed axis of rotation $\boldsymbol{e}_r$ and its derivatives are given by

$$\boldsymbol{\omega}(t) = \dot{s}_r(t)\boldsymbol{e}_r \tag{6.101a}$$

$$\dot{\boldsymbol{\omega}}(t) = \ddot{s}_r(t)\boldsymbol{e}_r \tag{6.101b}$$

$$\ddot{\boldsymbol{\omega}}(t) = s_r^{(3)}(t)\boldsymbol{e}_r \tag{6.101c}$$

$$\boldsymbol{\omega}^{(3)}(t) = s_r^{(4)}(t)\boldsymbol{e}_r \tag{6.101d}$$

In summary, the prescribed trajectory, that CABLEV has to follow, is specified by

$$\boldsymbol{x} = \begin{bmatrix} \boldsymbol{r}(t) \\ \boldsymbol{\varphi}(t) \end{bmatrix} \tag{6.102}$$

Then the servo constraints are used to prescribe the desired movement of the payload platform.

6.3.5 Application of index reduction by minimal extension

The equations of motion of CABLEV are composed of the kinematic differential Equation (6.60), the dynamic equations of the drive system (6.85), the dynamic equations of the platform (6.88), the holonomic constraints (6.68), the control constraint (6.81), and the control constraints in terms of the load coordinates of the platform (6.102).

The resulting index-5 DAEs assume the form

$$\dot{\boldsymbol{r}} = \boldsymbol{v} \tag{6.103a}$$

$$\dot{\boldsymbol{\varphi}} = \boldsymbol{H}_\omega(\boldsymbol{\varphi})\boldsymbol{\omega} \tag{6.103b}$$

$$\boldsymbol{M}_1\ddot{\boldsymbol{p}} = \boldsymbol{f}_1(\boldsymbol{p},\dot{\boldsymbol{p}}) + \boldsymbol{B}_1^T\boldsymbol{u} + \boldsymbol{G}_p^T(\boldsymbol{p},\boldsymbol{x})\boldsymbol{\lambda} \tag{6.103c}$$

$$\boldsymbol{M}_2\dot{\boldsymbol{t}} = \boldsymbol{f}_2 + \boldsymbol{G}_s^T(\boldsymbol{p},\boldsymbol{x})\boldsymbol{\lambda} \tag{6.103d}$$

$$\boldsymbol{\Phi}(\boldsymbol{p},\boldsymbol{x}) = \boldsymbol{0} \tag{6.103e}$$

$$\Phi_0(\boldsymbol{p}) = 0 \tag{6.103f}$$

$$\boldsymbol{r} = \gamma_r(t) \tag{6.103g}$$

$$\boldsymbol{\varphi} = \gamma_\varphi(t) \tag{6.103h}$$

The detailed description of equations of motion are given in Appendix A.6. As described in Chapter 5, the index reduction by minimal extension approach can be applied to the DAEs (6.103a)–(6.103h) as well.

The holonomic constraints (6.103e) are enforced by the Lagrange multipliers in Equation (6.103c), and thus are not responsible for the index 5 structure of the DAEs. Accordingly, the control constraints (6.103f) and (6.103g) should be differentiated twice with respect to time. The constraint conditions at the acceleration level are appended to the original DAEs (6.103a)–(6.103h).

In general, the control constraint (6.103h) should also be differentiated twice with respect to time. However, instead of the Bryant angles $\boldsymbol{\varphi}$, the angular velocity $\boldsymbol{\omega} \in \mathbb{R}^3$ of the rigid platform represented in $\mathcal{K}_0$ has been used to describe the rotational motion of the platform in Equation (6.103d). Therefore, the motion planning of the trajectory of the platform provides directly the prescribed angular velocity and acceleration as function of time.

Then the dummy derivatives are introduced to replace the corresponding derivatives of the coordinates. After the application of index reduction by minimal extension, the index-reduced DAEs are written in the following form

$$\begin{aligned}
M_1 \widetilde{p} &= f_1(p, \dot{p}) + B_1^T u + G_p^T(p, x)\lambda && \text{(6.104a)}\\
M_2 \widetilde{t} &= f_2 + G_s^T(p, x)\lambda && \text{(6.104b)}\\
\Phi(p, x) &= 0 && \text{(6.104c)}\\
\Phi_0(p) &= 0 && \text{(6.104d)}\\
\ddot{\Phi}_0(\widetilde{p}) &= 0 && \text{(6.104e)}\\
r &= \gamma_r(t) && \text{(6.104f)}\\
\varphi &= \gamma_\varphi(t) && \text{(6.104g)}\\
\widetilde{r} &= \ddot{\gamma}_r(t) && \text{(6.104h)}\\
\omega &= \gamma_\omega(t) && \text{(6.104i)}\\
\widetilde{\omega} &= \dot{\gamma}_\omega(t) && \text{(6.104j)}
\end{aligned}$$

Here, the vector $\widetilde{p}$ denotes the second derivative of the robot coordinates p,

$$\widetilde{p} = \begin{bmatrix} \ddot{p}_{g0} & \widetilde{p}_{g1} & \ddot{p}_{g2} & \ddot{p}_{g3} & \ddot{p}_{c1} & \ddot{p}_{c2} & \ddot{p}_{c3} \end{bmatrix}^T \tag{6.105}$$

where $\ddot{p}_{g1}$ is replaced by the dummy variable $\widetilde{p}_{g1}$. The remaining dummy variables are then defined by

$$\widetilde{r} = \ddot{r}, \quad \widetilde{\omega} = \dot{\omega}, \quad \widetilde{t} = \begin{bmatrix} \widetilde{r} \\ \widetilde{\omega} \end{bmatrix} \tag{6.106}$$

The constraint condition (6.104e) is represented by

$$\widetilde{p}_{g1} - \frac{1}{2}\left(\ddot{p}_{g2} + \ddot{g}_{g3}\right) = 0 \tag{6.107}$$

Note that the acceleration $\ddot{p}_{g2}$ or $\ddot{p}_{g3}$ could also be selected as an alternative derivative which would be replaced by the corresponding dummy variable, for example, $\widetilde{p}_{g2}$ or $\widetilde{p}_{g3}$. The resulting DAEs (6.104a)–(6.104j) have the index of 3 after the application of index reduction by minimal extension procedure. Similarly, the index-3 DAEs after index reduction are given in detail in Appendix A.6.

6.3.6 Differential flatness

The mechanical system under consideration can be classified as a differentially flat system, since all the state variables as well as control inputs, can be algebraically expressed in terms of the flat outputs $\gamma(t)$ and the time derivatives up to a certain order without integrating any differential equations [39, 53]. In particular, the system dynamics can be inverted according to

$$y = f_y(\gamma, \dot{\gamma}, \cdots, \gamma^{(\alpha-1)}) \tag{6.108a}$$

$$u = f_u(\gamma, \dot{\gamma}, \cdots, \gamma^{(\alpha)}) \tag{6.108b}$$

where α is a finite natural number by one smaller than the index of DAEs. The inversion of the system dynamics is realized by the generalized inverse kinematics and inverse dynamics as in [53].

Generalized inverse kinematics

The robot coordinates $\boldsymbol{p}$ as well as the Lagrange multipliers $\boldsymbol{\lambda}$ can be expressed in terms of the flat outputs $\boldsymbol{x}$ and the derivatives up to the second order, for example

$$\boldsymbol{p} = \boldsymbol{f}_p(\boldsymbol{x}, \dot{\boldsymbol{s}}, \ddot{\boldsymbol{s}}) \tag{6.109}$$

For this, the dynamic equations (6.104b), constraint equations (6.104c) and (6.104d) constitute a new set of differential-algebraic equations in the following form

$$\boldsymbol{M}_2 \, \ddot{\boldsymbol{s}} = \boldsymbol{f}_2(\boldsymbol{x}, \dot{\boldsymbol{s}}) + \boldsymbol{G}_s^T(\boldsymbol{p}, \boldsymbol{x})\boldsymbol{\lambda} \tag{6.110a}$$

$$\boldsymbol{\Phi}(\boldsymbol{p}, \boldsymbol{x}) = \boldsymbol{0} \tag{6.110b}$$

$$\Phi_0(\boldsymbol{p}) = 0 \tag{6.110c}$$

For the unknown robot coordinates $\boldsymbol{p} \in \mathbb{R}^7$ and Lagrange multipliers $\boldsymbol{\lambda} \in \mathbb{R}^3$, Newton's method can be applied to solve the above nonlinear equations (6.110a)–(6.110c). In this case, the generalized inverse kinematics is performed at the position level.

The robot velocities $\dot{\boldsymbol{p}}$ are expressed in terms of the flat outputs $\boldsymbol{x}$ and the derivatives up to the third order, such as

$$\dot{\boldsymbol{p}} = \boldsymbol{f}_{\dot{p}}(\boldsymbol{x}, \dot{\boldsymbol{s}}, \ddot{\boldsymbol{s}}, \boldsymbol{s}^{(3)}) \tag{6.111}$$

Differentiating Equation (6.110a)–(6.110c) with respect to time yields a set of linear equations for $\dot{\boldsymbol{p}}$ and $\dot{\boldsymbol{\lambda}}$,

$$\begin{bmatrix} \boldsymbol{L}_0 & \boldsymbol{G}_s^T \\ \boldsymbol{G}_p & \boldsymbol{0} \\ \boldsymbol{G}_{p0} & \boldsymbol{0} \end{bmatrix} \begin{bmatrix} \dot{\boldsymbol{p}} \\ \dot{\boldsymbol{\lambda}} \end{bmatrix} = \begin{bmatrix} \boldsymbol{M}_2 \boldsymbol{s}^{(3)} + \boldsymbol{L}_2 \ddot{\boldsymbol{s}} + \boldsymbol{L}_1 \dot{\boldsymbol{s}} \\ -\boldsymbol{G}_s \dot{\boldsymbol{s}} \\ 0 \end{bmatrix} \tag{6.112}$$

or in compact form

$$\boldsymbol{A}(\boldsymbol{x}, \boldsymbol{p}, \boldsymbol{\lambda}) \begin{bmatrix} \dot{\boldsymbol{p}} \\ \dot{\boldsymbol{\lambda}} \end{bmatrix} = \boldsymbol{f}_v \left(\boldsymbol{x}, \dot{\boldsymbol{s}}, \ddot{\boldsymbol{s}}, \boldsymbol{s}^{(3)}, \boldsymbol{p}, \boldsymbol{\lambda} \right) \tag{6.113}$$

with the terms

$$\boldsymbol{L}_0 = \frac{\partial (\boldsymbol{G}_s^T \boldsymbol{\lambda})}{\partial \boldsymbol{p}} \tag{6.114a}$$

$$\boldsymbol{L}_1 = \frac{\partial}{\partial \boldsymbol{x}} \left(\boldsymbol{M}_2 \ddot{\boldsymbol{s}} - \boldsymbol{f}_2 - \boldsymbol{G}_s^T \boldsymbol{\lambda} \right) \boldsymbol{H} \tag{6.114b}$$

$$\boldsymbol{L}_2 = -\frac{\partial \boldsymbol{f}_2}{\partial \dot{\boldsymbol{s}}} \tag{6.114c}$$

$$\boldsymbol{G}_{p0} = \frac{\partial \Phi_0}{\partial \boldsymbol{p}} = \begin{bmatrix} 0 & 1 & -\frac{1}{2} & -\frac{1}{2} & 0 & 0 & 0 \end{bmatrix} \tag{6.114d}$$

At the velocity level, the robot velocities $\dot{\boldsymbol{p}}$ and $\dot{\boldsymbol{\lambda}}$ can thus be obtained by solving Equation (6.112) via Newton's method numerically.

Analogously, the robot accelerations $\ddot{\boldsymbol{p}}$ are expressed in terms of the flat outputs $\boldsymbol{x}$ and the derivatives up to the fourth order,

$$\ddot{\boldsymbol{p}} = \boldsymbol{f}_{\ddot{p}}(\boldsymbol{x}, \dot{\boldsymbol{s}}, \ddot{\boldsymbol{s}}, \boldsymbol{s}^{(3)}, \boldsymbol{s}^{(4)}) \tag{6.115}$$

Differentiating Equation (6.113) with respect to time yields the linear equations for $\ddot{\boldsymbol{p}}$ and $\ddot{\boldsymbol{\lambda}}$ at the acceleration level,

$$\begin{bmatrix} \boldsymbol{L}_0 & \boldsymbol{G}_s^T \\ \boldsymbol{G}_p & \boldsymbol{0} \\ \boldsymbol{G}_{p0} & \boldsymbol{0} \end{bmatrix} \begin{bmatrix} \ddot{\boldsymbol{p}} \\ \ddot{\boldsymbol{\lambda}} \end{bmatrix} = \begin{bmatrix} \boldsymbol{M}_2 \boldsymbol{s}^{(4)} + \boldsymbol{L}_2 \boldsymbol{s}^{(3)} + (\boldsymbol{L}_1 + \dot{\boldsymbol{L}}_2) \ddot{\boldsymbol{s}} + \dot{\boldsymbol{L}}_1 \dot{\boldsymbol{s}} - \boldsymbol{L}_3 \\ -\boldsymbol{G}_s \ddot{\boldsymbol{s}} - \dot{\boldsymbol{G}}_s \dot{\boldsymbol{s}} - \dot{\boldsymbol{G}}_p \dot{\boldsymbol{p}} \\ 0 \end{bmatrix} \tag{6.116}$$

or in compact form

$$\boldsymbol{A}(\boldsymbol{x}, \boldsymbol{p}, \boldsymbol{\lambda}) \begin{bmatrix} \ddot{\boldsymbol{p}} \\ \ddot{\boldsymbol{\lambda}} \end{bmatrix} = \boldsymbol{f}_a \left(\boldsymbol{x}, \dot{\boldsymbol{s}}, \ddot{\boldsymbol{s}}, \boldsymbol{s}^{(3)}, \boldsymbol{s}^{(4)}, \boldsymbol{p}, \dot{\boldsymbol{p}}, \boldsymbol{\lambda}, \dot{\boldsymbol{\lambda}} \right) \tag{6.117}$$

with the terms

$$\dot{L}_2 = \frac{\partial L_2}{\partial \dot{s}} \ddot{s} \tag{6.118a}$$

$$\dot{L}_1 = \frac{\partial L_1}{\partial x} H\dot{s} + \frac{\partial L_1}{\partial p} \dot{p} + \frac{\partial L_1}{\partial \lambda} \dot{\lambda} \tag{6.118b}$$

$$L_3 = \dot{L}_0 \dot{p} + \dot{G}_s^T \dot{\lambda} \tag{6.118c}$$

$$\dot{L}_0 = \frac{\partial L_0}{\partial x} H\dot{s} + \frac{\partial L_0}{\partial \lambda} \dot{\lambda} \tag{6.118d}$$

$$\dot{G}_s^T = \frac{\partial G_s^T}{\partial x} H\dot{s} + \frac{\partial G_s^T}{\partial p} \dot{p} \tag{6.118e}$$

$$\dot{G}_p = \frac{\partial G_p}{\partial x} H\dot{s} + \frac{\partial G_p}{\partial p} \dot{p} \tag{6.118f}$$

Symbolic manipulations can be used to obtain the sets of equations (6.110), (6.112) and (6.117).

Inverse dynamics

Since the robot coordinates $\boldsymbol{p}$, velocities $\dot{\boldsymbol{p}}$ and accelerations $\ddot{\boldsymbol{p}}$ are already known, the control inputs $\boldsymbol{u}$ can be algebraically calculated by solving Equation (6.104c),

$$\boldsymbol{u} = (\boldsymbol{B}_1^T)^{-1} \left(\boldsymbol{M}_1 \ddot{\boldsymbol{p}} - \boldsymbol{f}_1(\boldsymbol{p}, \dot{\boldsymbol{p}}) - \boldsymbol{G}_p^T(\boldsymbol{p}, \boldsymbol{x}) \boldsymbol{\lambda} \right) \tag{6.119}$$

which are expressed in terms of the flat outputs $\boldsymbol{x}$ and their derivatives up to the fourth order,

$$\boldsymbol{u} = \boldsymbol{f}_u(\boldsymbol{x}, \dot{\boldsymbol{s}}, \ddot{\boldsymbol{s}}, \boldsymbol{s}^{(3)}, \boldsymbol{s}^{(4)}) \tag{6.120}$$

The fourth order derivative in the expression of control inputs $\boldsymbol{u}$ implies that $\alpha = 4$ and the index of DAEs is $\nu = \alpha + 1 = 5$. The inversion of the system dynamics of CABLEV naturally provides the feedforward control law by Equation (6.119) and proves that the mechanical system under consideration is differentially flat.

6.3.7 Numerical example

The numerical simulation makes use of the following parameters: $m_0 = 380\,\mathrm{kg}$, $m_1 = m_2 = m_3 = 35\,\mathrm{kg}$, $m = 12.5\,\mathrm{kg}$, $J_1 = J_2 = 0.75\,\mathrm{kg \cdot m^2}$, $J_3 = 1.5\,\mathrm{kg \cdot m^2}$, $r_1 = r_2 = r_3 = 0.1\,\mathrm{m}$, and $l_2 = 0.6\,\mathrm{m}$.

Different trajectories of translational or/and rotational motion of the playload platform are then investigated to present the simulation results.

Rotation about space-fixed vertical axis

The prescribed trajectory of rotational motion is generated as explained in Subsection 6.3.4 and there exists no translational motion in this case study. The initial orientation described by the Bryant angles is given by

$$\boldsymbol{\varphi}_a = \begin{bmatrix} 0 \\ 0 \\ 0 \end{bmatrix} \quad \text{at} \quad t = 0 \tag{6.121}$$

and the final orientation is given by

$$\boldsymbol{\varphi}_e = \begin{bmatrix} 0 \\ 0 \\ \frac{\pi}{3} \end{bmatrix} \quad \text{at} \quad t = 20\,\text{s} \tag{6.122}$$

The initial configuration of the robot system is specified by

$$\boldsymbol{p} = \begin{bmatrix} 1.2\,\text{m} & (1 + \frac{3}{5}\sqrt{3})\,\text{m} & 1\,\text{m} & 1\,\text{m} & 3\,\text{m} & 3\,\text{m} & 3\,\text{m} \end{bmatrix}^T \tag{6.123}$$

at $t_0 = 0$, and the initial platform coordinates are given by

$$\boldsymbol{x} = \begin{bmatrix} \boldsymbol{r} & \boldsymbol{\varphi} \end{bmatrix}^T = \begin{bmatrix} (1 + \frac{\sqrt{3}}{5})\,\text{m} & 1.2\,\text{m} & 3\,\text{m} & 0 & 0 & 0 \end{bmatrix}^T \tag{6.124}$$

Note that the position vector $\boldsymbol{r}$ keeps constant during the rotational motion. The numerical results for the time step size $\Delta t = 0.01\,\text{s}$ are presented in Fig. 6.21 and 6.22, in which the robot coordinates, control inputs and Lagrange multipliers are displayed. The simulated rotational motion of CABLEV is presented in Fig. 6.23 and 6.24 with some snapshots at consecutive points in time.

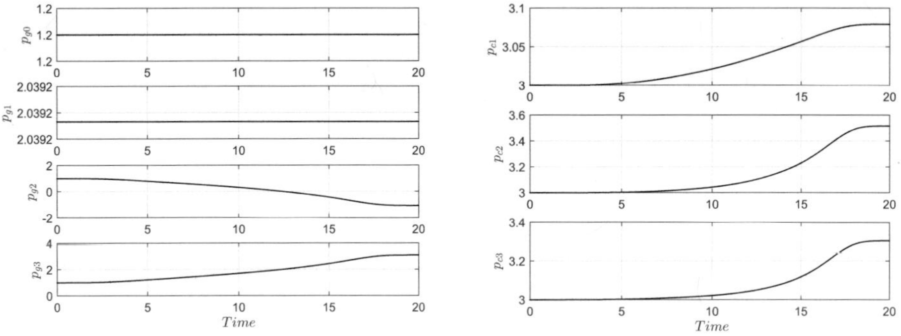

Figure 6.21: CABLEV: Numerical results of rotational motion with $\Delta t = 0.01\,\mathrm{s}$.

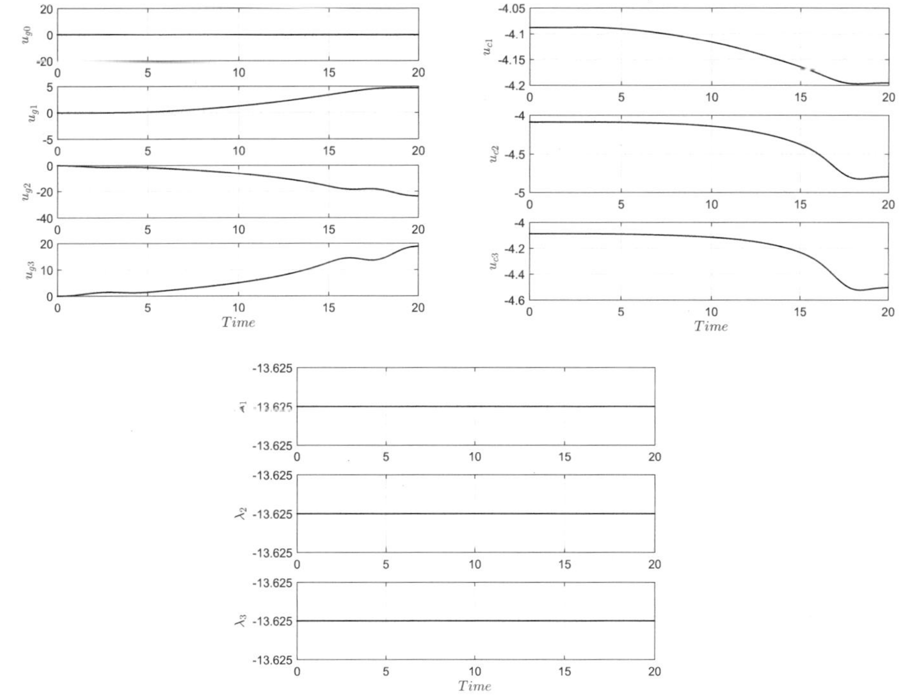

Figure 6.22: CABLEV: Numerical results of rotational motion with $\Delta t = 0.01\,\mathrm{s}$.

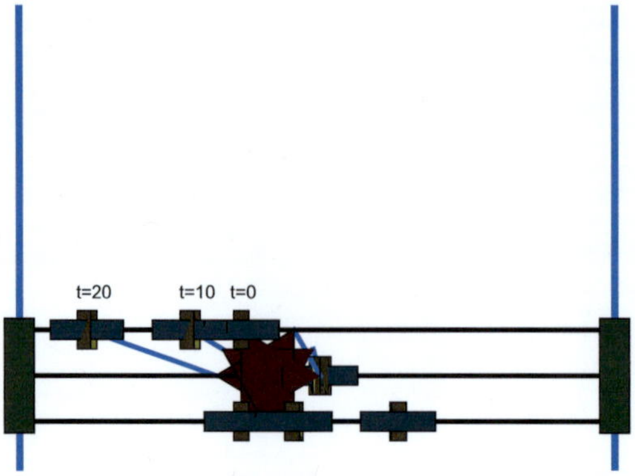

Figure 6.23: CABLEV: Snapshots of rotational motion at specific points in time.

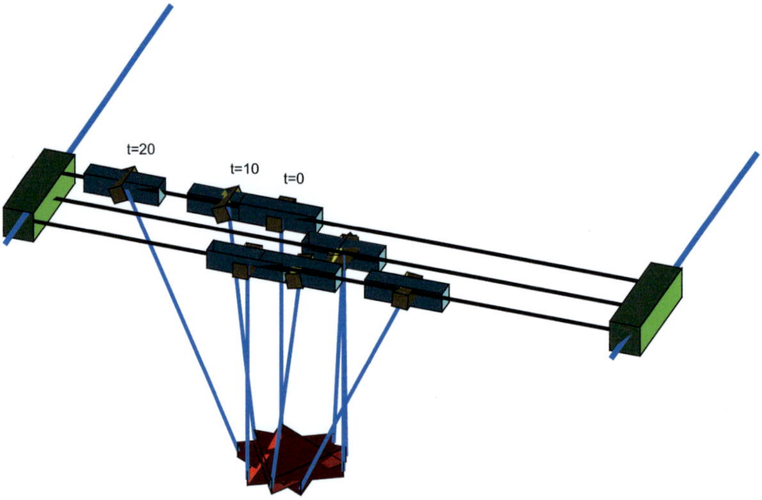

Figure 6.24: CABLEV: Snapshots of rotational motion at specific points in time.

Translation along a straight line

The prescribed trajectory of translational motion is generated in the same way as before and there exists no rotational motion in this case study. The initial position of the center of mass of platform is given by

$$\gamma_{r0} = \begin{bmatrix} (1 + \frac{\sqrt{3}}{5})\,\mathrm{m} \\ 1.2\,\mathrm{m} \\ 3\,\mathrm{m} \end{bmatrix} \quad \text{at} \quad t = 0 \tag{6.125}$$

and the final position is given by

$$\gamma_{rf} = \begin{bmatrix} (3 + \frac{\sqrt{3}}{5})\,\mathrm{m} \\ 5\,\mathrm{m} \\ 1\,\mathrm{m} \end{bmatrix} \quad \text{at} \quad t = 20\,\mathrm{s} \tag{6.126}$$

The initial configuration of the robot system is specified by

$$p = \begin{bmatrix} 1.2\,\mathrm{m} & (1 + \frac{3}{5}\sqrt{3})\,\mathrm{m} & 1\,\mathrm{m} & 1\,\mathrm{m} & 3\,\mathrm{m} & 3\,\mathrm{m} & 3\,\mathrm{m} \end{bmatrix}^T \tag{6.127}$$

at $t_0 = 0$, and the initial platform coordinates are given by

$$x = \begin{bmatrix} r & \varphi \end{bmatrix}^T = \begin{bmatrix} (1 + \frac{\sqrt{3}}{5})\,\mathrm{m} & 1.2\,\mathrm{m} & 3\,\mathrm{m} & 0 & 0 & 0 \end{bmatrix}^T \tag{6.128}$$

Note that the orientation φ of the platform keeps unchanged during the translational motion. The numerical results for the time step size $\Delta t = 0.01\,\mathrm{s}$ are shown in Fig. 6.25, in which the robot coordinates, control inputs and Lagrange multipliers are presented. The simulated translational motion of CABLEV is presented in Fig. 6.26 with some snapshots at consecutive points in time.

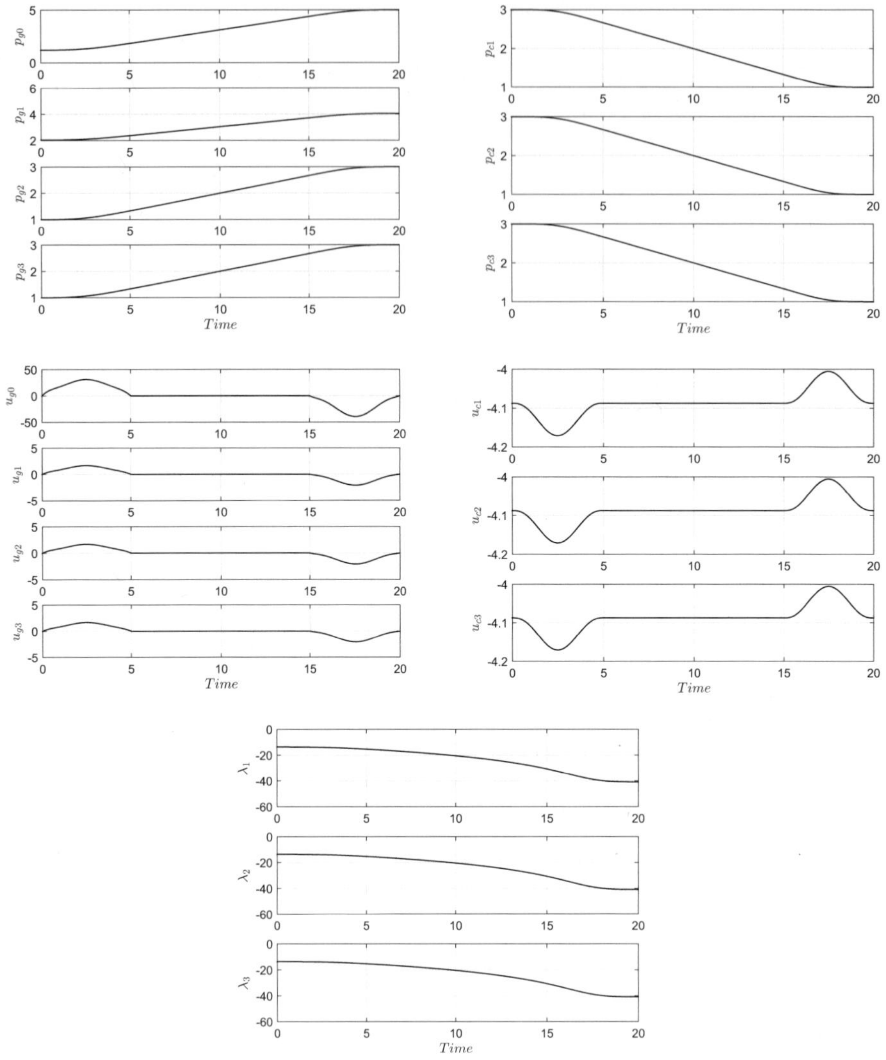

Figure 6.25: CABLEV: Numerical results of translational motion with $\Delta t = 0.01\,\text{s}$.

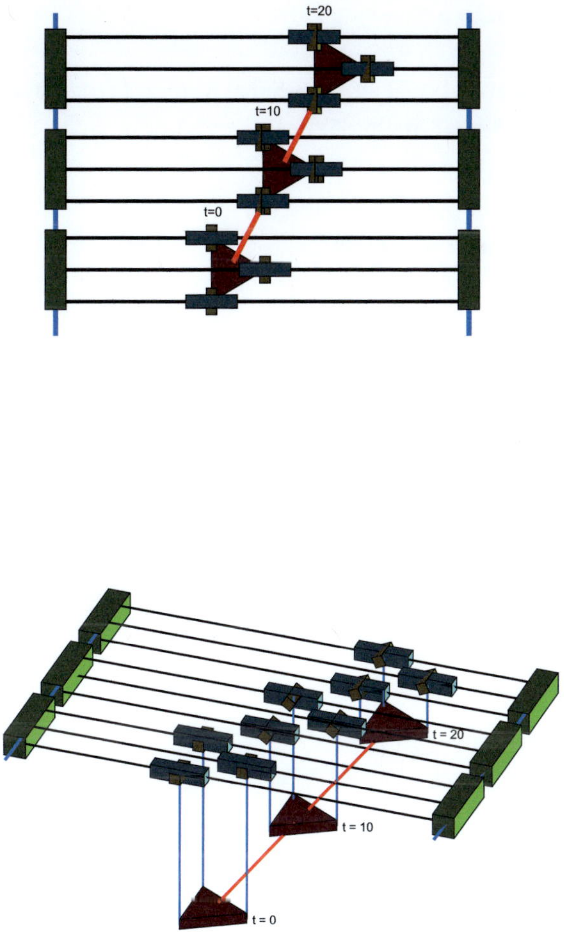

Figure 6.26: CABLEV: Snapshots of translational motion at specific points in time.

Translation with rotation

A combination of translational and rotational motion of the payload platform is investigated in this case study. The initial position and orientation of the payload platform are given by

$$\gamma_{r0} = \begin{bmatrix} (1+\frac{\sqrt{3}}{5})\,\mathrm{m} \\ 1.2\,\mathrm{m} \\ 3\,\mathrm{m} \end{bmatrix} \qquad \varphi_a = \begin{bmatrix} 0 \\ 0 \\ 0 \end{bmatrix} \quad \text{at} \quad t = 0 \tag{6.129}$$

and the final position and orientation are given by

$$\gamma_{rf} = \begin{bmatrix} (3+\frac{\sqrt{3}}{5})\,\mathrm{m} \\ 5\,\mathrm{m} \\ 1\,\mathrm{m} \end{bmatrix} \qquad \varphi_e = \begin{bmatrix} \pi/6 \\ \pi/4 \\ \pi/3 \end{bmatrix} \quad \text{at} \quad t = 20\,\mathrm{s} \tag{6.130}$$

The initial configuration of the robot system is specified by

$$p = \begin{bmatrix} 1.2\,\mathrm{m} & (1+\frac{3}{5}\sqrt{3})\,\mathrm{m} & 1\,\mathrm{m} & 1\,\mathrm{m} & 3\,\mathrm{m} & 3\,\mathrm{m} & 3\,\mathrm{m} \end{bmatrix}^T \tag{6.131}$$

at $t_0 = 0$, and the initial load coordinates are given by

$$x = \begin{bmatrix} r & \varphi \end{bmatrix}^T = \begin{bmatrix} (1+\frac{\sqrt{3}}{5})\,\mathrm{m} & 1.2\,\mathrm{m} & 3\,\mathrm{m} & 0 & 0 & 0 \end{bmatrix}^T \tag{6.132}$$

The numerical results for the time step size $\Delta t = 0.01\,\mathrm{s}$ are shown in Fig. 6.27 and 6.28, in which the robot coordinates, control inputs and Lagrange multipliers are presented. The simulated motion of CABLEV is presented in Fig. 6.29 and 6.30 with some snapshots at consecutive points in time.

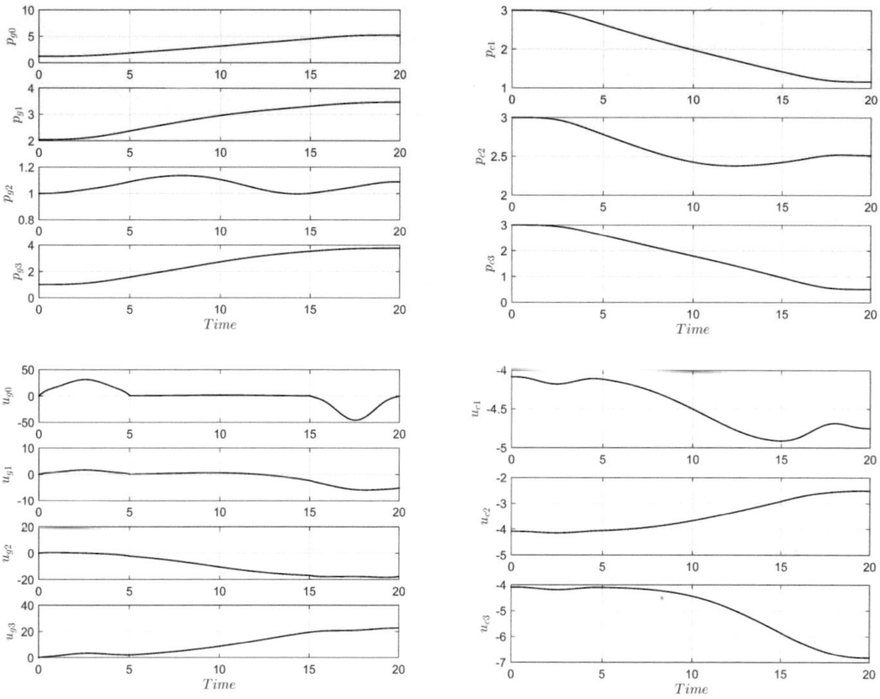

Figure 6.27: CABLEV: Numerical results obtained with $\Delta t = 0.01\,\mathrm{s}$.

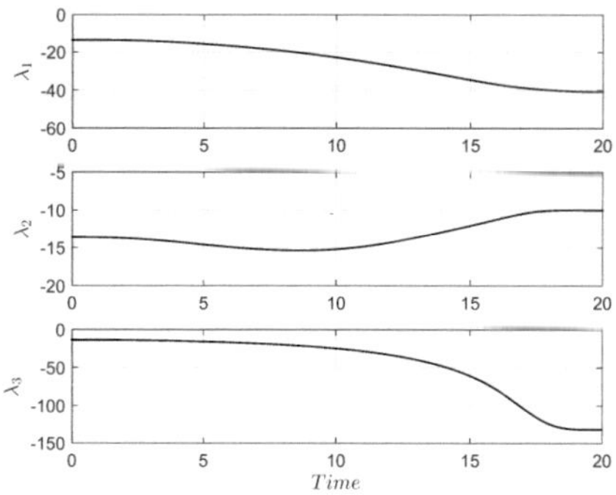

Figure 6.28: CABLEV: Numerical results obtained with $\Delta t = 0.01\,\mathrm{s}$.

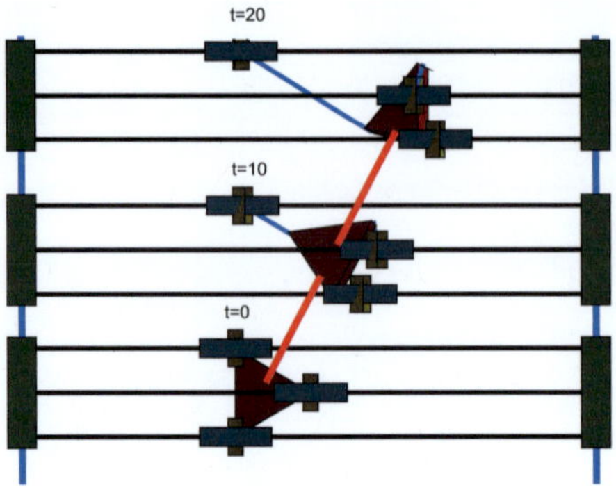

Figure 6.29: CABLEV: Snapshots of CABLEV at specific points in time.

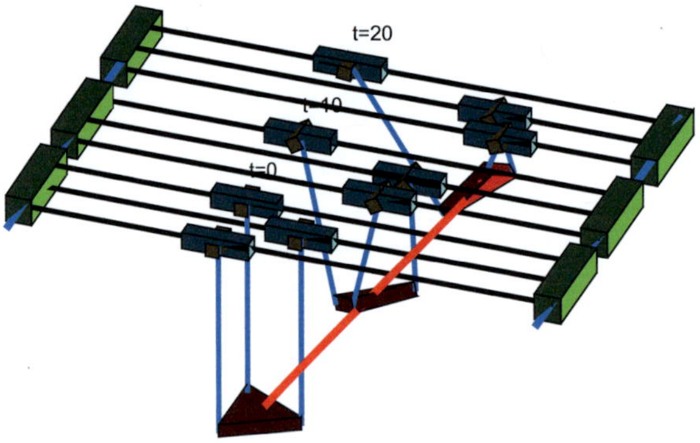

Figure 6.30: CABLEV: Snapshots of CABLEV at specific points in time.

7 Summary and outlook

7.1 Summary

This thesis deals with the inverse dynamics simulation of underactuated multibody systems. In particular, the focus is laid on differentially flat underactuated mechanical systems. The formulation of underactuated systems relies on the choice of coordinates. In the present thesis, minimal coordinates, dependent coordinates and redundant coordinates are used to formulate either the feedforward dynamics problems or the inverse dynamics problems for underactuated systems. In addition, the use of servo constraints provides an approach to the formulation of the inverse dynamics analysis. In this case numerical methods are needed to solve such servo constraint problems of underactuated systems, since the governing equations of motion of the system are in the form of DAEs with high index, for example, index five.

The projection method is investigated in depth for diverse formulations within two numerical examples. The numerical simulation results agree well with the reference analytical solution derived by using the property of differential flatness. The projection method requires the computation of projection matrices, which are constant Boolean-type in the case of using redundant coordinates and are time-dependent in the case of using minimal coordinates. The projection matrices for the redundant coordinates formulation are much simpler than those for the minimal coordinates formulation, since the use of redundant coordinats or dependent coordinates leads to some simplifications in the formulation of the problem. In addition, the redundant coordinates formulation is characterized by a constant mass matrix, whereas the minimal coordinates formulation leads to a complex configuration-dependent mass matrix. Special attention is thus paid to the formulation in terms of redundant coordinates and dependent coordinates. The projection method can yield an index reduction from five to three and can not be applied to systems with a singular mass matrix, such as the examples of US Navy cranes described in Chapter 6.

A newly proposed index reduction method, called index reduction by minimal extension, is developed in this work. It can be applied to solve servo constraint problems of underactuated mechanical systems. In the applications index reduction by minimal extension is also performed for different formulations of representative numerical examples, such as for the redundant coordinates formulation and the minimal coordinates formulation. The numerical simulation results are identical to the flatness-based solution in the case of very small time step size. The new approach can reduce the index from five to three and even to one if it is applied twice to the problem at hand. The resulting index-1 DAEs are purely algebraic and this shows that the underactuated system under consideration is differentially flat. Moreover, in the new approach it is not necessary to compute projection matrices. In addition, index reduction by minimal extension is applied to some advanced examples, such as the US Navy crane, for which the mass matrix is singular in the redundant coordinates formulation, and the undetermined cable suspension manipulator, for which the payload is modeled as a rigid body and the motion is thus much more complicated.

7.2 Outlook

In this thesis the backward Euler method is used as a time stepping scheme for the direct discretization of the resulting DAEs, whose index has been reduced by applying appropriate index reduction methods. Since the backward Euler method is only first order accurate, the design of energy consistent second order or higher order accurate schemes needs to be further considered and investigated in future work. For example, the dynamic behaviour of underactuated servo constraint problems is very sensitive to the application of the mid-point-type rule. Oscillations in the numerical results are often observed for very small time step size and this indicates that the integration scheme is not numerically stable.

The study in this work focuses on differentially flat underactuated multibody systems, which have no internal dynamics in the system. There exist also nonflat underactuated multibody systems such as manipulators with both passive and active joints [32] and the Blajer' car example [33, 82]. In the case of nonflat underactuated systems, the stability of the internal dynamics [82, 81, 26, 33] ensures the controllability of the system and thus is of paramount importance. The stability can be ensured through the design of the system's properties like the inertial value or the position of the center of percussion chosen as the control output.

The theory of a Cosserat point has been applied to the rotationless formulation of both rigid and flexible multibody dynamics. In the case of rigid multibody dynamics, the rotationless formulation yields a set of index-3 DAEs for constrained mechanical systems. Thus, the reduction of index from three to one may be investigated by application of index reduction by minimal extension in the context of the Cosserat point so that computation time can be saved further. In the case of flexible multibody dynamics, there exist also many underactuated multibody systems such as manipulators with flexible members. The solution of trajectory tracking control problems of such systems needs to be developed by applying index reduction by minimal extension.

A Detailed explanations

A.1 Gâteaux derivative

Using Gâteaux derivative, Equation (3.11) yields

$$\begin{aligned}\delta S_d &= \sum_{n=0}^{N-1} \delta L_d(\boldsymbol{q}_n, \boldsymbol{q}_{n+1}) = \sum_{n=0}^{N-1} \frac{d}{d\varepsilon}\bigg|_{\varepsilon=0} L_d(\boldsymbol{q}_n + \varepsilon\delta\boldsymbol{q}_n, \boldsymbol{q}_{n+1} + \varepsilon\delta\boldsymbol{q}_{n+1}) \\ &= \sum_{n=0}^{N-1} \left[D_1 L_d(\boldsymbol{q}_n, \boldsymbol{q}_{n+1}) \cdot \delta\boldsymbol{q}_n + D_2 L_d(\boldsymbol{q}_n, \boldsymbol{q}_{n+1}) \cdot \delta\boldsymbol{q}_{n+1}\right]\end{aligned} \tag{A.1}$$

A.2 Planar US Navy crane with neglected mass

The equations of motion are given in detail as follows:

$$\begin{aligned}
\frac{J_1}{r_1^2}\ddot{L}_1 &= -\lambda_1 L_1 + \frac{u_1}{r_1} \\
\frac{J_2}{r_2^2}\ddot{L}_2 &= \lambda_3(L_0 - L_2) + \frac{u_2}{r_2} \\
0 &= -\lambda_2 L_0 + \lambda_3(L_2 - L_0) \\
0 &= \lambda_1(x_0 + l\sin\alpha) + \lambda_2 x_0 + \lambda_3(x_0 - x) \\
0 &= \lambda_1(z_0 + l\cos\alpha) + \lambda_2 z_0 + \lambda_3(z_0 - z) \\
m\ddot{x} &= \lambda_3(x - x_0) \\
m\ddot{z} &= \lambda_3(z - z_0) - mg \\
0 &= \frac{1}{2}((x_0 + l\ \sin\alpha)^2 + (z_0 + l\ \cos\alpha)^2 - L_1^2) \\
0 &= \frac{1}{2}(x_0^2 + z_0^2 - L_0^2) \\
0 &= \frac{1}{2}((x - x_0)^2 + (z - z_0)^2 - (L_2 - L_0)^2) \\
x &= \gamma_1(t) \\
z &= \gamma_2(t)
\end{aligned} \tag{A.2}$$

A.3 Planar US Navy crane with nonzero mass

The equations of motion are given in detail as follows:

$$
\begin{aligned}
\frac{J_1}{r_1^2}\ddot{L}_1 &= -\lambda_1 L_1 + \frac{u_1}{r_1} \\
\frac{J_2}{r_2^2}\ddot{L}_2 &= \lambda_4(L_0 - L_2) + \frac{u_2}{r_2} \\
\frac{J_3}{r_3^2}\ddot{L}_3 &= -\lambda_3 L_3 + \frac{u_3}{r_3} \\
0 &= -\lambda_2 L_0 + \lambda_4(L_2 - L_0) \\
m_0\ddot{x}_0 &= \lambda_1(x_0 + (l+s)\sin\alpha) + \lambda_2(x_0 + s\sin\alpha) + \lambda_3 x_0 + \lambda_4(x_0 - x) \\
m_0\ddot{z}_0 &= \lambda_1(z_0 + (l+s)\cos\alpha) + \lambda_2(z_0 + s\cos\alpha) + \lambda_3 z_0 + \lambda_4(z_0 - z) - m_0 g \\
m\ddot{x} &= \lambda_4(x - x_0) \\
m\ddot{z} &= \lambda_4(z - z_0) - mg \\
0 &= \frac{1}{2}((x_0 + (l+s)\sin\alpha)^2 + (z_0 + (l+s)\cos\alpha)^2 - L_1^2) \\
0 &= \frac{1}{2}((x_0 + s\,\sin\alpha)^2 + (z_0 + s\,\cos\alpha)^2 - L_0^2) \\
0 &= \frac{1}{2}(x_0^2 + z_0^2 - L_3^2) \\
0 &= \frac{1}{2}((x - x_0)^2 + (z - z_0)^2 - (L_2 - L_0)^2) \\
x &= \gamma_1(t) \\
z &= \gamma_2(t) \\
z_0 &= \gamma_3(t)
\end{aligned}
\tag{A.3}
$$

A.4 3D US Navy crane with neglected mass

The equations of motion are given in detail as follows:

$$
\begin{aligned}
\frac{J_1}{r_1^2}\ddot{L}_1 &= -\lambda_1 L_1 + \frac{u_1}{r_1} \\
\frac{J_3}{r_3^2}\ddot{L}_3 &= \lambda_4(L_0 - L_3) + \frac{u_3}{r_3} \\
0 &= -\lambda_2 L_0 + \lambda_4(L_3 - L_0) \\
0 &= \lambda_1(x_0 - \beta_1 x_3) + \lambda_2(x_0 - x_3) + \lambda_4(x_0 - x) \\
0 &= \lambda_1(y_0 - \beta_1 y_3) + \lambda_2(y_0 - y_3) + \lambda_4(y_0 - y) \\
0 &= \lambda_1(z_0 - \beta_1 z_3) + \lambda_2(z_0 - z_3) + \lambda_4(z_0 - z) \\
M\ddot{x}_3 &= -\lambda_1\beta_1(x_0 - \beta_1 x_3) + \lambda_2(x_3 - x_0) + \lambda_3 x_3 - u_4\frac{y_3}{r^2} \\
M\ddot{y}_3 &= -\lambda_1\beta_1(y_0 - \beta_1 y_3) + \lambda_2(y_3 - y_0) + \lambda_3 y_3 + u_4\frac{x_3}{r^2} \\
m\ddot{x} &= \lambda_4(x - x_0) \\
m\ddot{y} &= \lambda_4(y - y_0) \\
m\ddot{z} &= \lambda_4(z - z_0) - mg \\
0 &= \frac{1}{2}((x_0 - \beta_1 x_3)^2 + (y_0 - \beta_1 y_3)^2 + (z_0 - \beta_1 z_3)^2 - L_1^2) \\
0 &= \frac{1}{2}((x_0 - x_3)^2 + (y_0 - y_3)^2 + (z_0 - z_3)^2 - L_0^2) \\
0 &= \frac{1}{2}(x_3^2 + y_3^2 - r^2) \\
0 &= \frac{1}{2}((x - x_0)^2 + (y - y_0)^2 + (z - z_0)^2 - (L_3 - L_0)^2) \\
x &= \gamma_1(t) \\
y &= \gamma_2(t) \\
z &= \gamma_3(t)
\end{aligned}
\tag{A.4}
$$

A.5 3D US Navy crane with nonzero mass

The equations of motion are given in detail as follows:

$$
\begin{aligned}
\frac{J_1}{r_1^2}\ddot{L}_1 &= -\lambda_2 L_1 + \frac{u_1}{r_1} \\
\frac{J_2}{r_2^2}\ddot{L}_2 &= -\lambda_3 L_2 + \frac{u_2}{r_2} \\
\frac{J_3}{r_3^2}\ddot{L}_3 &= \lambda_5 (L_0 - L_3) + \frac{u_3}{r_3} \\
0 &= -\lambda_4 L_0 + \lambda_5 (L_3 - L_0) \\
m_0 \ddot{x}_0 &= \lambda_2 (x_0 - \beta_1 x_3) + \lambda_3 (x_0 - \beta_2 x_3) + \lambda_4 (x_0 - x_3) + \lambda_5 (x_0 - x) \\
m_0 \ddot{y}_0 &= \lambda_2 (y_0 - \beta_1 y_3) + \lambda_3 (y_0 - \beta_2 y_3) + \lambda_4 (y_0 - y_3) + \lambda_5 (y_0 - y) \\
m_0 \ddot{z}_0 &= \lambda_2 (z_0 - \beta_1 z_3) + \lambda_3 (z_0 - \beta_2 z_3) + \lambda_4 (z_0 - z_3) + \lambda_5 (z_0 - z) - m_0 g \\
M \ddot{x}_3 &= \lambda_1 x_3 - \lambda_2 \beta_1 (x_0 - \beta_1 x_3) - \lambda_3 \beta_2 (x_0 - \beta_2 x_3) - \lambda_4 (x_0 - x_3) - u_4 \frac{y_3}{r^2} \\
M \ddot{y}_3 &= \lambda_1 y_3 - \lambda_2 \beta_1 (y_0 - \beta_1 y_3) - \lambda_3 \beta_2 (y_0 - \beta_2 y_3) - \lambda_4 (y_0 - y_3) + u_4 \frac{x_3}{r^2} \\
m \ddot{x} &= \lambda_5 (x - x_0) \\
m \ddot{y} &= \lambda_5 (y - y_0) \\
m \ddot{z} &= \lambda_5 (z - z_0) - mg \\
0 &= \frac{1}{2}(x_3^2 + y_3^2 - r^2) \\
0 &= \frac{1}{2}((x_0 - \beta_1 x_3)^2 + (y_0 - \beta_1 y_3)^2 + (z_0 - \beta_1 z_3)^2 - L_1^2) \\
0 &= \frac{1}{2}((x_0 - \beta_2 x_3)^2 + (y_0 - \beta_2 y_3)^2 + (z_0 - \beta_2 z_3)^2 - L_2^2) \\
0 &= \frac{1}{2}((x_0 - x_3)^2 + (y_0 - y_3)^2 + (z_0 - z_3)^2 - L_0^2) \\
0 &= \frac{1}{2}((x - x_0)^2 + (y - y_0)^2 + (z - z_0)^2 - (L_3 - L_0)^2) \\
x &= \gamma_1(t) \\
y &= \gamma_2(t) \\
z &= \gamma_3(t)
\end{aligned}
\tag{A.5}
$$

A.6 Cable suspension manipulator

The index-5 DAEs of CABLEV model assume the form:

$$
\begin{aligned}
\begin{bmatrix} \dot{r}_x \\ \dot{r}_y \\ \dot{r}_z \end{bmatrix} &= \begin{bmatrix} v_x \\ v_y \\ v_z \end{bmatrix} \\
\begin{bmatrix} \dot{\varphi}_1 \\ \dot{\varphi}_2 \\ \dot{\varphi}_3 \end{bmatrix} &= \frac{1}{\cos\varphi_2} \begin{bmatrix} \cos\varphi_2 & \sin\varphi_1 \sin\varphi_2 & -\cos\varphi_1 \sin\varphi_2 \\ 0 & \cos\varphi_1 \cos\varphi_2 & \sin\varphi_1 \cos\varphi_2 \\ 0 & -\sin\varphi_1 & \cos\varphi_1 \end{bmatrix} \begin{bmatrix} \omega_x \\ \omega_y \\ \omega_z \end{bmatrix} \\
\boldsymbol{M}_1 \begin{bmatrix} \ddot{p}_{g0} \\ \ddot{p}_{g1} \\ \ddot{p}_{g2} \\ \ddot{p}_{g3} \\ \ddot{p}_{c1} \\ \ddot{p}_{c2} \\ \ddot{p}_{c3} \end{bmatrix} &= \begin{bmatrix} 1 & & & & & & \\ & 1 & & & & & \\ & & 1 & & & & \\ & & & 1 & & & \\ & & & & \frac{1}{r_1} & & \\ & & & & & \frac{1}{r_2} & \\ & & & & & & \frac{1}{r_3} \end{bmatrix} \begin{bmatrix} u_{g0} \\ u_{g1} \\ u_{g2} \\ u_{g3} \\ u_{c1} \\ u_{c2} \\ u_{c3} \end{bmatrix} + \begin{bmatrix} -\boldsymbol{c}_1^T \boldsymbol{e}_y & -\boldsymbol{c}_2^T \boldsymbol{e}_y & -\boldsymbol{c}_3^T \boldsymbol{e}_y \\ -\boldsymbol{c}_1^T \boldsymbol{e}_x & 0 & 0 \\ 0 & -\boldsymbol{c}_2^T \boldsymbol{e}_x & 0 \\ 0 & 0 & -\boldsymbol{c}_3^T \boldsymbol{e}_x \\ -p_{c1} & 0 & 0 \\ 0 & -p_{c2} & 0 \\ 0 & 0 & -p_{c3} \end{bmatrix} \begin{bmatrix} \lambda_1 \\ \lambda_2 \\ \lambda_3 \end{bmatrix} \\
\boldsymbol{M}_2 \begin{bmatrix} \dot{v}_x \\ \dot{v}_y \\ \dot{v}_z \\ \dot{\omega}_x \\ \dot{\omega}_y \\ \dot{\omega}_z \end{bmatrix} &= \begin{bmatrix} 0 \\ 0 \\ mg \\ -(J_x - J_y)\omega_y \omega_z \\ -(J_x - J_z)\omega_z \omega_x \\ -(J_y - J_x)\omega_x \omega_y \end{bmatrix} + \begin{bmatrix} \boldsymbol{c}_1 & \boldsymbol{c}_2 & \boldsymbol{c}_3 \\ -\tilde{\boldsymbol{d}}_1^{\,T} \boldsymbol{c}_1 & -\tilde{\boldsymbol{d}}_2^{\,T} \boldsymbol{c}_2 & -\tilde{\boldsymbol{d}}_3^{\,T} \boldsymbol{c}_3 \end{bmatrix} \begin{bmatrix} \lambda_1 \\ \lambda_2 \\ \lambda_3 \end{bmatrix} \\
0 &= \frac{1}{2}(\boldsymbol{c}_1 \cdot \boldsymbol{c}_1 - p_{c1}^2) \\
0 &= \frac{1}{2}(\boldsymbol{c}_2 \cdot \boldsymbol{c}_2 - p_{c2}^2) \\
0 &= \frac{1}{2}(\boldsymbol{c}_3 \cdot \boldsymbol{c}_3 - p_{c3}^2) \\
0 &= p_{g1} - \frac{1}{2}(p_{g2} + p_{g3}) - \sqrt{3} l_2 \\
r_x &= \gamma_x \\
r_y &= \gamma_y \\
r_z &= \gamma_z \\
\varphi_1 &= \gamma_1 \\
\varphi_2 &= \gamma_2 \\
\varphi_3 &= \gamma_3
\end{aligned}
\qquad (A.6)
$$

After application of index reduction by minimal extension, the index-3 DAEs of CABLEV model assume the form

$$
\begin{aligned}
\boldsymbol{M}_1 \begin{bmatrix} \ddot{p}_{g0} \\ \overline{p}_{g1} \\ \ddot{p}_{g2} \\ \ddot{p}_{g3} \\ \ddot{p}_{c1} \\ \ddot{p}_{c2} \\ \ddot{p}_{c3} \end{bmatrix} &= \begin{bmatrix} 1 & & & & & & \\ & 1 & & & & & \\ & & 1 & & & & \\ & & & 1 & & & \\ & & & & \frac{1}{r_1} & & \\ & & & & & \frac{1}{r_2} & \\ & & & & & & \frac{1}{r_3} \end{bmatrix} \begin{bmatrix} u_{g0} \\ u_{g1} \\ u_{g2} \\ u_{g3} \\ u_{c1} \\ u_{c2} \\ u_{c3} \end{bmatrix} + \begin{bmatrix} -\boldsymbol{c}_1^T \boldsymbol{e}_y & -\boldsymbol{c}_2^T \boldsymbol{e}_y & -\boldsymbol{c}_3^T \boldsymbol{e}_y \\ -\boldsymbol{c}_1^T \boldsymbol{e}_x & 0 & 0 \\ 0 & -\boldsymbol{c}_2^T \boldsymbol{e}_x & 0 \\ 0 & 0 & -\boldsymbol{c}_3^T \boldsymbol{e}_x \\ -p_{c1} & 0 & 0 \\ 0 & -p_{c2} & 0 \\ 0 & 0 & -p_{c3} \end{bmatrix} \begin{bmatrix} \lambda_1 \\ \lambda_2 \\ \lambda_3 \end{bmatrix} \\
\boldsymbol{M}_2 \begin{bmatrix} \overline{r}_x \\ \overline{r}_y \\ \overline{r}_z \\ \overline{z}_x \\ \overline{z}_y \\ \overline{z}_z \end{bmatrix} &= \begin{bmatrix} 0 \\ 0 \\ mg \\ -(J_x - J_y)\omega_y\omega_z \\ -(J_x - J_z)\omega_z\omega_x \\ -(J_y - J_x)\omega_x\omega_y \end{bmatrix} + \begin{bmatrix} \boldsymbol{c}_1 & \boldsymbol{c}_2 & \boldsymbol{c}_3 \\ -\tilde{\boldsymbol{d}}_1^T \boldsymbol{c}_1 & -\tilde{\boldsymbol{d}}_2^T \boldsymbol{c}_2 & -\tilde{\boldsymbol{d}}_3^T \boldsymbol{c}_3 \end{bmatrix} \begin{bmatrix} \lambda_1 \\ \lambda_2 \\ \lambda_3 \end{bmatrix} \\
0 &= \frac{1}{2}(\boldsymbol{c}_1 \cdot \boldsymbol{c}_1 - p_{c1}^2) \\
0 &= \frac{1}{2}(\boldsymbol{c}_2 \cdot \boldsymbol{c}_2 - p_{c2}^2) \\
0 &= \frac{1}{2}(\boldsymbol{c}_3 \cdot \boldsymbol{c}_3 - p_{c3}^2) \\
0 &= p_{g1} - \frac{1}{2}(p_{g2} + p_{g3}) - \sqrt{3} l_2 \\
0 &= \overline{p}_{g1} - \frac{1}{2}(\ddot{p}_{g2} + \ddot{p}_{g3}) \\
r_x &= \gamma_x \\
r_y &= \gamma_y \\
r_z &= \gamma_z \\
\overline{r}_x &= \ddot{\gamma}_x \\
\overline{r}_y &= \ddot{\gamma}_y \\
\overline{r}_z &= \ddot{\gamma}_z \\
\varphi_1 &= \gamma_1 \\
\varphi_2 &= \gamma_2 \\
\varphi_3 &= \gamma_3 \\
\boldsymbol{\omega}(t) &= \boldsymbol{e}_r \dot{s}_r(t) \\
\overline{\boldsymbol{z}} &= \boldsymbol{e}_r \ddot{s}_r(t)
\end{aligned}
\tag{A.7}
$$

Bibliography

[1] R. Altmann. Index reduction for operator differential-algebraic equations in elastodynamics. *Z. Angew. Math. Mech. (ZAMM)*, 93(9):648–664, 2013. doi: 10.1002/zamm.201200125.

[2] R. Altmann, P. Betsch, and Y. Yang. Index reduction by minimal extension for the inverse dynamics simulation of cranes. *Multibody System Dynamics*, 36(3): 295–321, 2016.

[3] J. Angeles. *Fundamentals of robotic mechanical systems*. Springer-Verlag, 2nd edition, 2003.

[4] M. Arnold and O. Brüls. Convergence of the generalized-α scheme for constrained mechanical systems. *Multibody System Dynamics*, 18(2):185–202, 2007.

[5] H. Aschemann. *Optimale Trajektorienplanung sowie modellgestützte Steuerung und Regelung für einen Brückenkran*. Fortschritt-Berichte VDI, Reihe 8: Meß-, Steuerungs- und Regelungstechnik, Nr. 929, Düsseldorf, Germany, 2002.

[6] U.M. Ascher and L.R. Petzold. *Computer methods for ordinary differential equations and differential-algeraic equations*. SIAM, 1998.

[7] O.A. Bauchau and C.L. Bottasso. On the design of energy preserving and decaying schemes for flexible, nonlinear multi-body systems. *Comput. Methods Appl. Mech. Engrg.*, 169:61–79, 1999.

[8] O.A. Bauchau, C.L. Bottasso, and L. Trainelli. Robust integration schemes for flexible multibody systems. *Comput. Methods Appl. Mech. Engrg.*, 192:395–420, 2003.

[9] P. Betsch. The discrete null space method for the energy consistent integration of constrained mechanical systems. Part I: Holonomic constraints. *Comput. Methods Appl. Mech. Engrg.*, 194(50-52):5159–5190, 2006.

[10] P. Betsch. Skriptum Technische Mechanik. 2011.

[11] P. Betsch. Lecture notes in Numerische Strukturmechanik. 2015.

[12] P. Betsch and C. Hesch. Energy-momentum conserving schemes for frictionless dynamic contact problems. Part I: NTS method. In P. Wriggers and U. Nackenhorst, editors, *IUTAM Symposium on Computational Methods in Contact Mechanics*, volume 3, pages 77–79, Berlin, 2007. Springer-Verlag.

[13] P. Betsch and S. Leyendecker. The discrete null space method for the energy consistent integration of constrained mechanical systems. Part II: Multibody dynamics. *Int. J. Numer. Methods Eng.*, 67(4):499–552, 2006.

[14] P. Betsch and N. Sänger. On the use of geometrically exact shells in a conserving framework for flexible multibody dynamics. *Computer Methods in Applied Mechanics and Engineering*, 198:1609–1630, 2009.

[15] P. Betsch and N. Sänger. A nonlinear finite element framework for flexible multibody dynamics: Rotationless formulation and energy-momentum conserving discretization. In Carlo L. Bottasso, editor, *Multibody Dynamics*, volume 12 of *Computational Methods and Applications*, pages 119–141. Springer-Verlag, 2009.

[16] P. Betsch and R. Siebert. Rigid body dynamics in terms of quaternions: Hamiltonian formulation and conserving numerical integration. *Int. J. Numer. Methods Eng.*, 79(4):444–473, 2009. doi: 10.1002/nme.2586.

[17] P. Betsch and P. Steinmann. Constrained integration of rigid body dynamics. *Computer Methods in Applied Mechanics and Engineering*, 191:467–488, 2001.

[18] P. Betsch and P. Steinmann. Conservation properties of a time fe method. Part III: Mechanical systems with holonomic constraints. *Int. J. Numer. Methods Eng.*, 53:2271–2304, 2002.

[19] P. Betsch and S. Uhlar. Energy-momentum conserving integration of multibody dynamics. *Multibody System Dynamics*, 17(4):243–289, 2007.

[20] P. Betsch, S. Uhlar, and M. Quasem. On the incorporation of servo constraints into a rotationless formulation of flexible multibody dynamics. In P. Masarati C.L. Bottasso and L. Trainelli, editors, *Proceedings of the ECCOMAS Thematic Conference on Multibody Dynamics*, Politecnico di Milano, Milano, Italy, 25-28 June 2007.

[21] P. Betsch, S. Uhlar, and M. Quasem. Numerical integration of discrete mechanical systems with mixed holonomic and control constraints. In *Proceedings of the 4th Asian Conference on Multibody Dynamics*, Jeju, Korea, 2008.

[22] P. Betsch, C. Hesch, N. Sänger, and S. Uhlar. Variational integrators and energy-momentum schemes for flexible multibody dynamics. *J. Comput. Nonlinear Dynam.*, 5(3):031001, 2010.

[23] P. Betsch, R. Siebert, and N. Sänger. Natural coordinates in the optimal control of multibody systems. *Journal of Computational and Nonlinear Dynamics*, 7(1):011009, 2012.

[24] P. Betsch, R. Altmann, and Y. Yang. Numerical integration of underactuated mechanical systems subjected to mixed holonomic and servo constraints. In J.M. Font-Llagunes, editor, *Multibody Dynamics*, volume 42 of *Computational Methods in Applied Sciences*, chapter 1, pages 1–18. Springer-Verlag, 2016.

[25] W. Blajer. Dynamics and control of mechanical systems in partly specified motion. *J. Franklin Inst.*, 334B:407–426, 1997.

[26] W. Blajer. The use of servo-constraints in the inverse dynamics analysis of underactuated multibody systems. *Transactions of the ASME, Journal of Computational and Nonlinear Dynamics*, 9(4):041008/1–11, accepted: October 2013. doi: 10.1115/1.4025855.

[27] W. Blajer and K. Kolodziejczyk. A geometric approach to solving problems of control constraints: Theory and a DAE framework. *Multibody Syst. Dyn.*, 11: 343–364, 2004.

[28] W. Blajer and K. Kolodziejczyk. A computational framework for control design of rotary cranes. In *Proceedings of the ECCOMAS Thematic Conference on Advances in computational Multibody Dynamics*, Madrid, Spain, June 21-24 2005.

[29] W. Blajer and K. Kolodziejczyk. Control of underactuated mechanical systems with servo-constraints. *Nonlinear Dyn.*, 50:781–791, 2007.

[30] W. Blajer and K. Kolodziejczyk. Dependent versus independent variable formulation for the dynamics and control of cranes. *Solid State Phenomena*, 147-149: 221–230, 2009.

[31] W. Blajer and K. Kołodziejczyk. Improved DAE formulation for inverse dynamics simulation of cranes. *Multibody Syst. Dyn.*, 25(2):131–143, 2011. doi: 10.1007/s11044-010-9227-6.

[32] W. Blajer and K. Kolodziejczyk. Controllability of planar manipulators with passive joints subject to servo-constraints. In *Proceedins of the 3rd Joint International*

Conference on Multibody System Dynamics and the 7th Asian Conference on Multibody Dynamics (IMSD-ACMD), Busan, Korea, June 30-July 3 2014.

[33] W. Blajer, R Seifried, and K. Kołodziejczyk. Diversity of servo-constraint problems for underactuated mechanical systems: A case study illustration. *Solid State Phenomena*, 198:473–482, 2013, doi:10.4028/www.scientific.net/SSP.198.473. Online at `http://www.scientific.net`.

[34] K. E. Brenan, S. L. Campbell, and L. R. Petzold. *Numerical solution of initial-value problems in differential-algebraic equations*. Society for Industrial and Applied Mathematics (SIAM), Philadelphia, PA, 1996. ISBN 0-89871-353-6.

[35] K.E. Brenan, S.L. Campbell, and L.R. Petzold. Numerical solution of initial-value problems in differential-algebra equations. *SIAM*, second ed., 1996.

[36] JJ. Craig. *Introduction to robotics: mechanics and control*. Pearson/Prentice Hall, Upper Saddle River, 3rd edition, 2003.

[37] J. García de Jalón. Twenty-five years of natural coordinates. *Multibody Syst. Dyn.*, 18(1):15–33, 2007.

[38] J. García de Jalón and J. Bayo. *Kinematic and Dynamic Simulation of Multibody Systems: The Real-Time Challenge*. Springer-Verlag, Berlin, 2009.

[39] M. Fliess, J. Lévine, P. Martin, and P. Rouchou. Flatness and defect of nonlinear systems: Introductory theory and examples. *International Journal of Control*, 61: 1327–1361, 1995.

[40] M. Franke. *Discretisation techniques for large deformation computational contact elastodynamics*. Phd thesis, Karlsruhe Institute of Technology, 2014.

[41] M. Franke, C. Hesch, and P. Betsch. An augmentation technique for large deformation frictional contact problems. *Int. J. Numer. Methods Eng.*, 94:513–534, 2013.

[42] M. Géradin and A. Cardona. *Flexible Multibody Dynamics - A Finite Element Approach*. Wiley, 2001.

[43] H. Goldstein. *Classical Mechanics*. Addison-Wesley Publishing Company, 1980.

[44] O. Gonzalez. Time integration and discrete hamiltonian systems. *J. Nonlinear Sci.*, 6:449–467, 1996.

[45] O. Gonzalez. Mechanical systems subject to holonomic constraints: Differential-algebraic formulations and conservative integration. *Physica D*, 132:165–174, 1999.

[46] D. Gross, W. Hauger, J. Schröder, W. A. Wall, and S. Govindjee. *Engineering Mechanics 3*. Springer, Berlin, Heidelberg, 2014.

[47] M. Groß. *Conserving Time Integrators for Nonlinear Elastodynamics*. Phd thesis, University of Kaiserslautern, 2004.

[48] M. Groß. *Higher-order accurate and energy-momentum consistent discretisation of dynamic finite deformation thermo-viscoelasticity*. Habilitation thesis, University of Siegen, 2009.

[49] M. Groß and P. Betsch. Galerkin-based energy-momentum consistent time-stepping algorithms for classical nonlinear thermo-elastodynamics. *Mathematics and Computers in Simulation*, 82(4):718–770, 2011.

[50] E. Hairer and G. Wanner. *Solving Ordinary Differential Equations II: Stiff and Differential-Algebraic Problems*. Springer, 1991.

[51] C. Hesch. *Mechanische Integratoren für Kontaktvorgänge deformierbarer Körper unter großen Verzerrungen*. Phd thesis, University of Siegen, 2007.

[52] T. Heyden. *Bahnregelung eines seilgeführten Handhabungssystems mit kinematisch unbestimmter Lastführung. Fortschritt-Bericht VDI, Reihe 8, Nr. 1100*. VDI-Verlag, Düsseldorf, 2006.

[53] T. Heyden and C. Woernle. Dynamics and flatness-based control of a kinematically undetermined cable suspension manipulator. *Multibody System Dynamics*, 16(2):155–177, 2006.

[54] K. P. Jankowski. *Inverse Dynamics Control in Rototics Applications*. Trafford Publishing, Victoria, BC, Canada, 2004.

[55] N. Khude and D. Negrut. A MATLAB implementation of the seven-body mechanism for implicit integration of the constrained equations of motion. Technical Report TR-2007-07, Simulation-Based Engineering Laboratory, The University of Wisconsin-Madison, 2007.

[56] V.I. Kirgetov. The motion of controlled mechanical systems with prescribed constraints (servoconstraints). *J. Appl. Maths. Mechs.*, 31(3):465–477, 1967.

[57] B. Kiss. *Motion planning and control of a class of flat and Liouvillian mechanical systems*. Phd thesis, Mines Paris Tech, 2001.

[58] B. Kiss, J. Lévine, and Ph. Müllhaupt. Modeling, flatness and simulation of a class of cranes. *Electrical Engineering*, 43(3):215–225, 1999.

[59] B. Kiss, J. Lévine, and Ph. Müllhaupt. Modeling and motion planning for a class of weight handling equipments. In *Proceedings of the 14th International Conference on Systems Engineering*, Conventry, UK, September 2000.

[60] M. Krüger. *Energie-Entropie-konsistente Zeitintegratoren für die nichtlineare Thermoviskoelastodynamik*. Phd thesis, University of Siegen, 2013.

[61] M. Krüger, M. Groß, and P. Betsch. A comparison of structure-preserving integrators for discrete thermoelastic systems. *Computational Mechanics*, 47(6):701–722, 2011.

[62] P. Kunkel and V. Mehrmann. Index reduction for differential-algebraic equations by minimal extension. *Z. Angew. Math. Mech. (ZAMM)*, 84(9):579–597, 2004. ISSN 1521-4001. doi: 10.1002/zamm.200310127.

[63] P. Kunkel and V. Mehrmann. *Differential-Algebraic Equations: Analysis and Numerical Solution*. European Mathematical Society, Zürich, 2006. doi: 10.4171/017.

[64] F. Kuypers. *Klassische Mechanik*. Willey, 2010.

[65] J. Lévine, P. Rouchou, G. Yuan, C. Grebogi, B.R. Hunt, E. Kostelich, E. Ott, and J. Yorke. On the control of us navy cranes. In *Proceedings of the European Control Conference*, pages 213–217, Brussels, Belgium, July 1997.

[66] A. Lew, J.E. Marsden, M. Ortiz, and M. West. Variational time integrators. *International Journal for Numerical Methods in Engineering*, 60:153–212, 2004.

[67] S. Leyendecker, P. Betsch, and P. Steinmann. The discrete null space method for the energy consistent integration of constrained mechanical systems. Part III: Flexible multibody dynamics. *Multibody System Dynamics*, 19(1-2):45–72, 2008.

[68] S. Leyendecker, J. E. Marsden, and M. Ortiz. Variational integrators for constrained dynamical systems. *Z. Angwe. Math. Mech.*, 88(9):677–708, 2008.

[69] A. De Luca. *Trajectory Control of Flexible Manipulators*. Springer, London, 1998.

[70] T. Maier. *Bahnsteuerung eines seilgeführten Handhabungssystems: Modellbildung, Simulation und Experiment. Fortschritt-Bericht VDI, Reihe 8, Nr. 1047*. VDI-Verlag, Düsseldorf, 2004.

[71] J. Marsden and M. West. Discrete mechanics and variational integrators. *Acta Numerica*, pages 357–514, 2001.

[72] J.E. Marsden and T.S. Ratiu. *Introduction to Mechanics and Symmetry*. Springer Verlag, 2nd edition, 1999.

[73] S. Mattsson and G. Söderlind. Index reduction in differential-algebraic equations using dummy derivatives. *SIAM J. Sci. Comput.*, 14:677–692, 1993.

[74] L. Meirovitch. *Methods of Analytical Dynamics*. McGraw-Hill Book Company, 1970.

[75] RM. Murray, Z. Li, and Sastry SS. *A mathematical introduction to robotic manipulators*. CRC Press, New York, 1994.

[76] A. Rosen. Applying the lagrange method to solve problems of control constraints. *Journal of Applied Mechanics*, 66:1013–1015, 1999.

[77] R.M. Rosenberg. *Analytical dynamics of discrete systems*. Plenum Press, 1977.

[78] P. Rouchon. Flatness based control of oscillators. *Z. Angew. Math. Mech. (ZAMM)*, 85(6):411–421, 2005.

[79] N. Sänger. *Elemente für die Dynamik flexibler Mehrkörpersysteme*. Phd thesis, University of Siegen, 2011.

[80] W.O. Schiehlen. *Multibody systems handbook*. Springer, 1990.

[81] R. Seifried. *Dynamics of Underactuated Multibody Systems: Modeling, Control and Optimal Design*. Springer, 2014.

[82] R. Seifried and W. Blajer. Analysis of servo-constraint problems for underactuated multibody systems. *Mechanical Sciences*, 4(1):113–129, 2013. doi: 10.5194/ms-4-113-2013.

[83] A. A. Shabana. *Dynamics of Multibody Systems*. Cambridge University Press, Cambridge, 2010.

[84] R. Siebert. *Mechanical integrators for the optimal control in multibody dynamics*. Phd thesis, University of Siegen, 2012.

[85] J.C. Simo and N. Tarnow. The discrete energy-momentum method. conserving algorithms for nonlinear elastodynamics. *Z. Angew. Math. Phys. (ZAMP)*, 43:757–792, 1992.

[86] M. Spong, S. Hutchinson, and M. Vidyasagar. *Robot modeling and control*. John Wiley, New York, 2006.

[87] S. Uhlar. *Energy consistent time-integration of hybrid multibody systems*. Phd thesis, University of Siegen, 2009.

[88] S. Uhlar and P. Betsch. On the rotationless formulation of multibody dynamics and its conserving numerical integration. In *Proceedings of the ECCOMAS Thematic Conference on Advances in computational Multibody Dynamics*, Milano, Italy, June 25-28 2007.

[89] S. Uhlar and P. Betsch. A rotationless formulation of multibody dynamics: Modeling of screw joints and incorporation of control constraints. *Multibody System Dynamics*, 22(1):69–95, 2009. doi: 10.1007/s11044-009-9149-3.

[90] M.J. van Nieuwstadt and R.M. Murray. Real-time trajectory generation for differentially flat systems. *International Journal of Robust and Nonlinear Control*, 8(11): 995–1020, 1998.

[91] J. von Schwerin. MultiBody System SIMulation. Springer, 1999.

[92] C. Woernle. *Mehrkörpersysteme, Eine Einführung in die Kinematik und Dynamik von Systemen starrer Körper*. Springer Verlag, Berlin Heidelberg, 2011.

[93] Y. Yang and P. Betsch. On the choice of coordinates for computational multibody dynamics. *Proc. Appl. Math. Mech. (PAMM)*, 11:75–76, 2011.

[94] Y. Yang and P. Betsch. Alternative approaches to the incorporation of control constraints in multibody dynamics. In *Proc. of the ECCOMAS Thematic Conference on Multibody Dynamics*, Zagreb, Croatia, July 1-4 2013.

[95] Y. Yang, P. Betsch, and R. Altmann. A numerical method for the servo constraint problem of underactuated mechanical systems. *Proc. Appl. Math. Mech. (PAMM)*, 15:79–80, 2015.

[96] Y. Yang, P. Betsch, and R. Altmann. A new approach to the incorporation of servo constraints in multibody dynamics. *Proc. Appl. Math. Mech. (PAMM)*, 16: 67–68, 2016.

Schriftenreihe des Instituts für Mechanik
ISSN 2363-4936

Hrsg. von: Prof. Dr.-Ing. habil. Peter Betsch, Prof. Dr.-Ing. habil. Thomas Seelig

---

Band 1 Marlon Franke
Discretisation techniques for large deformation computational contact elastodynamics. 2014
ISBN 978-3-7315-0278-4

Band 2 Philipp Hempel
Constitutive modeling of amorphous thermoplastic polymers with special emphasis on manufacturing processes. 2016
ISBN 978-3-7315-0550-1

Band 3 Martin Helbig
Mehrskalenmodellierung von Schädigung in gummimodifizierten thermoplastischen Kunststoffen. 2017
ISBN 978-3-7315-0571-6

Band 4 Maik Dittmann
Isogeometric analysis and hierarchical refinement for multi-field contact problems. 2017
ISBN 978-3-7315-0616-4

Band 5 Yinping Yang
Numerical methods for the inverse dynamics simulation of underactuated mechanical systems. 2017
ISBN 978-3-7315-0626-3

Die Bände sind unter www.ksp.kit.edu als PDF frei verfügbar oder als Druckausgabe bestellbar.